MOLECULAR
BIOLOGY
INTELLIGENCE
UNIT

SIGNAL TRANSDUCTION BY G PROTEIN-COUPLED RECEPTORS

BIOENERGETICS AND G PROTEIN ACTIVATION: PROTON TRANSFER AND GTP SYNTHESIS TO EXPLAIN THE EXPERIMENTAL FINDINGS

Paul H.J. Nederkoorn, Ph.D.
Henk Timmerman, Ph.D.
Gabriëlle M. Donné-Op den Kelder, Ph.D.
Leiden/Amsterdam Center for Drug Research
Amsterdam, The Netherlands

CHAPMAN & HALL
I(T)P An International Thomson Publishing Company

New York • Albany • Bonn • Boston • Cincinnati • Detroit • London • Madrid • Melbourne •
Mexico City • Pacific Grove • Paris • San Francisco • Singapore • Tokyo • Toronto • Washington

R.G. LANDES COMPANY
AUSTIN

MOLECULAR BIOLOGY INTELLIGENCE UNIT

SIGNAL TRANSDUCTION BY G PROTEIN-COUPLED RECEPTORS:
BIOENERGETICS AND G PROTEIN ACTIVATION: PROTON TRANSFER AND GTP SYNTHESIS TO EXPLAIN THE EXPERIMENTAL FINDINGS

R.G. LANDES COMPANY
Austin, Texas, U.S.A.

Softcover reprint of the hardcover 1st edition 1997

Please address all inquiries to the Publishers:
R.G. Landes Company, 810 South Church Street, Georgetown, Texas, U.S.A. 78626
Phone: 512/ 863 7762; FAX: 512/ 863 0081

North American distributor:
Chapman & Hall, 115 Fifth Avenue, New York, New York, U.S.A. 10003

CHAPMAN & HALL

U.S. and Canada **ISBN 978-1-4684-1409-7** **ISBN 978-1-4684-1407-3**
DOI 10.1007/ 978-1-4684-1407-3

Library of Congress Cataloging-in-Publication Data

Nederkoorn, Paul H.J.

Signal transduction by G protein-coupled receptors: bioenergetics and G protein activation: proton transfer and GTP synthesis to explain the experimental findings / Paul H.J. Nederkoorn, Henk Timmerman, and Gabrielle M. Donne-Op den Kelder.

p. cm. — (Molecular biology intelligence unit.)

Includes bibliographical references.

1. G Proteins. 2. Bioenergetics. 3. Cellular signal transduction. 4. Guanosine triphosphate—Synthesis. I. Timmerman, Henk. II. Donne-Op den Kelder, Gabrielle M. III. Title. IV. Series.

[DNLM: 1. Signal Transduction. 2. Receptors, Cell Surface. 3. G Proteins. 4. Energy Metabolism. QU 55 N371s 1996]

572'.643—dc21

DNLM/DLC

for Library of Congress 96-47443

Publisher's Note

R.G. Landes Company publishes six book series: *Medical Intelligence Unit, Molecular Biology Intelligence Unit, Neuroscience Intelligence Unit, Tissue Engineering Intelligence Unit, Biotechnology Intelligence Unit* and *Environmental Intelligence Unit.* The authors of our books are acknowledged leaders in their fields and the topics are unique. Almost without exception, no other similar books exist on these topics.

Our goal is to publish books in important and rapidly changing areas of bioscience and environment for sophisticated researchers and clinicians. To achieve this goal, we have accelerated our publishing program to conform to the fast pace in which information grows in bioscience. Most of our books are published within 90 to 120 days of receipt of the manuscript. We would like to thank our readers for their continuing interest and welcome any comments or suggestions they may have for future books.

Shyamali Ghosh
Publications Director
R.G. Landes Company

CONTENTS

ABBREVIATIONS

AA:	arachidonic acid
ADP:	adenosine 5'-diphosphate
ATP:	adenosine 5'-triphosphate
BR:	bacteriorhodopsin
cAMP:	cyclic adenosine 5'-monophosphate
CCCP:	carbonyl cyanide *m*-chlorophenylhydrazone
CMS:	Cytosensor Microphysiometer System
DAG:	diacylglycerol
DCCD:	*N,N'*-dicyclohexylcarbodiimide
ΔG_p:	Gibbs phosphorylation potential
$\Delta \mu_{H^+}$:	electrochemical proton gradient
Δp:	protonmotive force
$\Delta \psi$:	electrical membrane potential
E:	effector protein
E_h:	hydrolysis conformation of ATP synthase
E_s:	synthesis conformation of ATP synthase
ESE:	electrochemically stored energy
FCCP:	carbonyl cyanide *p*-trifluoromethoxyphenylhydrazone
G:	G protein
G_α:	α-subunit of G protein (G_β, G_γ idem)
G_α*:	activated G_α
GAP:	GTPase activating protein
GDP:	guanosine 5'-diphosphate
GDPβS:	guanosine 5'-[β-thio]diphosphate
GEF:	guanine nucleotide exchange factor
(c-)GMP:	(cyclic-) guanosine 5'-monophosphate
GPCR:	G protein-coupled receptor
G protein:	guanine nucleotide-binding regulatory protein
GTP:	guanosine 5'-triphosphate
GTPβS:	guanosine 5'-[β-thio]triphosphate
GTPγS:	guanosine 5'-[γ-thio]triphosphate

ABBREVIATIONS

H:	ligand (hormone)
HBC:	hydrogen-bonded chain
HL:	human leukemia
5-HT:	5-hydroxytryptamine (serotonine)
IP_3:	inositol triphosphate
MEP:	molecular electrostatic potential
NAD^+:	nicotinamide adenine dinucleotide
NADH:	reduced nicotinamide adenine dinucleotide
$NADP^+$:	nicotinamide adenine dinucleotide phosphate
NDP:	nucleoside diphosphate
NDPK:	nucleoside diphosphate kinase
N^{π}:	proximal tautomer
N-phase:	negative side of a membrane from which protons are pumped by a primary pump
N^{τ}:	tele tautomer
NTP:	nucleoside triphosphate
oGTP(γS):	2'-3'-dialdehyde GTP(γS) analogue
P_i:	inorganic phosphate
p[NH]ppG:	guanosine 5'-[β,γ-imido]triphosphate
P-phase:	positive side of a membrane to which protons are pumped by a primary pump
R:	receptor
R*:	activated receptor
ROS:	rod outer segment
SR-I:	phototaxis receptor sensory rhodopsin
(E)TCM:	(extended) ternary complex model
TM:	transmembrane domain
T-r:	transducin
$T\text{-}r_{\alpha}(^*)$:	(activated) G_{α} subunit of transducin
UDP:	uridine diphosphate

PREFACE

Receptors mediate ligand actions. This process starts with the binding of a ligand to a receptor on (or in) a target cell. Subsequently, a signal is transduced to another macromolecule that, in turn, triggers a cascade of effects such as alterations in enzyme activities, changes in gene expression and influences on ion channels. G protein-coupled receptors (GPCRs) are membrane-associated proteins, which—upon ligand binding—activate a G protein, thereby generating a diffusable intracellular signal. Part of this signal is formed by the so-called second messenger, including alterations in the concentrations of intracellular cAMP, cGMP, inositol triphosphate (IP_3) and cations (K^+, Na^+, Ca^{2+}). Hence, the G protein functions as a mediator between the receptor and the effector system regulating the second messenger production. Therefore, GPCRs and G proteins play a key role in transmembrane signalling. GPCRs respond to a wide variety of ligands, ranging from biogenic amines like histamine and adrenaline, odors and photons (in case of olfactory receptors and opsins, respectively), peptides (such as angiotensin II, bradykinin, tachykinins, endothelins, chemotactic and neuropeptides) to large glycoprotein hormones, such as the luteinizing and parathyroid hormone. Since numerous vital physiological (dys)functions are regulated by GPCR action, a better understanding of ligand binding and the subsequent processes through which the signalling pathways are influenced is of utmost importance for the development of therapeutical means against (wide-spread) diseases such as hypertension, asthma, allergic reactions, duodenal ulcers, Alzheimer's and Parkinson's diseases, schizophrenia, paranoia and depressions, cholera and pertussin (both caused by exogenous bacterial toxins), drug abuse, several deficiency syndromes, and adrenal, cortical and ovarian tumor growth.

Bioenergetic theories describe the mechanisms by which energy, made available by the oxidation of substrates or by the absorption of light, is used for energy-consuming reactions such as the synthesis of ATP. In this book a meeting is arranged between bioenergetics and G protein activation. Its purpose is two-fold: (i) bioenergetical mechanisms and G protein activation processes are reviewed and (ii) new viewpoints to well-established theories are offered, which aim at explaining all experimental findings, hopefully stimulating a series of experiments, and in this way contributing to the understanding of the molecular mechanisms underlying transmembrane signalling.

Recent theoretical studies[1,2] show that a ligand-mediated proton transfer, triggering signal transduction over a membrane, is theoretically feasible for a series of G protein-coupled amine receptors. Activation by an agonist ends up with one or more proton, per agonist bound, delivered intracellularly. By combining these findings with the deletion model for the origin of receptors proposed by Topiol,[3] an amine receptor can be seen as a proton pump, which lacks one piece in its proton shuttle, a hydrogen-bonded chain, to be able to pump continuously. After binding of the agonist, its pumping function is restored fully.

This concept stimulated us to consider that such a pumping mechanism may be of general relevance and we were struck by analogies found in the field of bioenergetics. A résumé of this field is given in the first part of this book. Crucial to bioenergetical processes are proton transfers. These protons are conducted via so-called primary pumps (pumping protons against an H^+ gradient) to secondary pumps (consuming the H^+ gradient generated by the primary pumps), establishing a so-called proton conductance circuit. Depending on the biological system under consideration, primary pumps differ greatly, whereas secondary pumps are highly conserved. In general, the proton translocating secondary pump is the ubiquitous $F_1{\cdot}F_0$-ATP synthase, which synthesizes ATP from ADP and P_i by pumping protons across a membrane with the aid of a proton gradient established by a primary pump such as bacteriorhodopsin. For related surveys and reviews see refs. 4-13.

In Part I, mechanisms by which protons are translocated for bioenergetical purposes are summarized. There are two alternative theories describing how translocated protons are possibly used for overall ATP synthesis: (i) the *direct* mechanism in which the synthesis of ATP requires protons to be delivered at a catalytic site, resulting in protonation of phosphate oxygens,[14,15] and (ii) the *indirect* mechanism in which translocated protons are thought to induce conformational changes in the $F_1{\cdot}F_0^-$ ATP synthase, resulting in the release of the ATP nucleotide.[5,9] In our view, these two contradicting theories can be reconciled when considering the concept of hydrogen-bonded chains acting as semiconducting proton wires.[16,17] The first part should be seen as an introduction to Part II.

In the second part, we focus on the mechanism of activation for both the ternary receptor complex (ligand-GPCR-G protein) and the G proteins themselves. Established models and their possible deficiencies are reviewed. Recently, we challenged current models and, based upon analogies found within the field of bioenergetics, we developed a new theory concerning the signal transduction via GPCRs and subsequent G protein activation.[18] The principles of our model will be described within the context of current ideas and insights and the experimental data on which our model is based. We will further develop its consequences and indicate avenues for further work.

Since proton wires play an essential role in our model, proton transfers proposed to occur upon ligand-activated GPCR activation will be outlined. The concept of Timms et al[1,2] is put into a broader context when we compare the

GPCRs in general with the ATP synthase system; membrane-bound GPCRs strongly resemble the F_0 unit functionally (not structurally). The ternary complex, i.e., hormone-GPCR-G protein (HR*G), translocates protons from the outside to the inside of a cell. The G protein in its turn can use these protons to generate GTP from GDP and P_i and thus resembles the F_1 unit of the ATP synthase in a functional manner. The G protein acts now as a GTP synthase. Remarkably, F_1 starts to hydrolyze ATP when it is decoupled from the F_0 unit; a decoupled $G_\alpha^* \cdot$ GTP entity also hydrolyzes GTP to GDP. Based upon analogies found in the field of bioenergetics (Part I), it becomes possible to explain the usual large drop in acid-extractable GDP observed in the period immediately following GPCR (rhodopsin) stimulation, whereas GTP levels remain about constant.[19] In contrast, current models for G protein activation start with an exchange of GTP for GDP, which should rather cause an increase in GDP and a drop in GTP levels.

Besides GDP/GTP exchange reactions, phosphorylation reactions have been reported for G proteins yielding a.o. GTP from GDP and P_i. The necessary phosphate transfers are found to proceed via G protein β subunits and are regulated by agonist-activated receptors.[20-22] These findings strongly support our theory of the ternary complex, and specifically the G protein, being a GTP synthase. The phosphorylation reactions are also proposed to result in an acceleration and amplification of the receptor mediated signal.[22]

In the mid-1960s, a well-known statement in the field of bioenergetics was "anyone who is not thoroughly confused just doesn't understand the problem."[9] In the field of signal transduction via GPCRs the uncertainty was seemingly resolved by the hypothesis of conformational changes to explain G protein activation (e.g., refs. 23-25). We postulate that there is more to it.

References

1. Timms D, Wilkinson AJ, Kelly DR et al. Interactions of Tyr^{377} in a ligand-activation model of signal transmission through β_1-adrenoceptor α-helices. Int J Quant Chem: Quant Biol Symp 1992; 19:197-215.
2. Timms D, Wilkinson AJ, Kelly DR et al. Ligand-activated trans-membrane proton transfer in β_1-adrenergic and m_2-muscarinergic receptors. Receptors and Channels 1994; 2:107-119.
3. Topiol S. The deletion model for the origin of receptors. Trends Biochem Sci 1987; 12:419-421.
4. Abrahams JP, Leslie AGW, Lutter R et al. Structure at 2.8 Å resolution of F_1-ATPase from bovine heart mitochondria. Nature 1994; 370:621-628.
5. Boyer PD. A perspective of the binding change mechanism for ATP synthesis. FASEB J 1989; 3:2164-2178.
6. Dencher NA, Büldt G, Heberle J et al. Light-triggered opening and closing of a hydrophobic gate controls vectorial proton transfer across bacteriorhodopsin. NATO ASI Ser, Ser B 1992; 291:171-185.

7. Henderson R, Baldwin JM, Ceska TA et al. Model for the structure of bacteriorhodopsin based on high-resolution electron cryo-microscopy. J Mol Biol 1990; 213:899-929.
8. Lanyi JK. Bacteriorhodopsin as a model for proton pumps. Nature 1995; 375:461-463.
9. Nicholls DG, Ferguson SJ. In: Bioenergetics 2. London:Academic Press, 1992.
10. Pedersen PL, Amzel LM. ATP synthases. Structure, reaction center, mechanism, and regulation of nature's most unique machines. J Biol Chem 1993; 268:9937-9940.
11. Penefsky HS, Cross RL. Structure and mechanism of F_0F_1-type ATP synthases and ATPases. Adv Enzymol 1991; 64:173-214.
12. Skulachev VP. Chemiosmotic concept of membrane bioenergetics: What is already clear and what is still waiting for elucidation? J Bioenerg Biomembr 1994; 26:589-598.
13. Tonomura Y. F_1-ATPase. In: Energy-transducing ATPases—Structure and Kinetics. Avon: Cambridge University Press, 1986:141-183.
14. Mitchell P. A chemiosmotic molecular mechanism for proton-translocating adenosine triphosphatases. FEBS Lett 1974; 43:189-194.
15. Mitchell P. Biochemical mechanism of protonmotivated phosphorylation in $F_0{\cdot}F_1$ adenosine triphosphate molecules. In: Lee CP, Schatze G, Dallner G, eds. Mitochondria and Microsomes. Reading: Addison Wesley, 1981:427-457.
16. Heberle J, Riesle J, Thiedemann G et al. Proton migation along the membrane surface and retarded surface to bulk transfer. Nature 1994; 370:379-382.
17. Morowitz HJ. Proton semiconductors and energy transduction in biological systems. Am J Physiol 1978; 235:R99-R114.
18. Nederkoorn PHJ, Timmerman H, Donné-Op den Kelder GM. Does the ternary complex act as a secondary proton pump and a GTP synthase? Trends Pharmacol Sci 1995; 16:156-161.
19. Robinson WE, Hagins WA. GTP hydrolysis in intact rod outer segments and the transmitter cycle in visual excitation. Nature 1979; 280:398-400.
20. Kaldenberg-Stasch S, Baden M, Fesseler B et al. Receptor-stimulated guanine nucleotide triphosphate binding to guanine nucleotide-binding regulatory proteins. Eur J Biochem 1994; 221:25-33.
21. Wieland T, Nürnberg B, Ulibarri I et al. Guanine nucleotide-specific phosphate transfer by guanine nucleotide-binding regulatory protein β-subuits. J Biol Chem 1993; 268:18111-18118.
22. Wieland T, Kaldenberg-Stasch S, Fesseler B et al. Regulation of G protein function by phosphorylation. Can J Physiol Pharmacol 1994; 72:S5.
23. Oliveira L, Paiva ACM, Sander C et al. A common step for signal transduction in G protein-coupled receptors. Trends Pharmacol Sci 1994; 15:170-172.
24. Samama P, Cotecchia S, Costa T et al. A mutation-induced activated state of the β_2-adrenergic receptor. Extending the ternary complex model. J Biol Chem 1993; 268:4625-4636.
25. Weinstein H. Computational simulations of molecular structure, dynamics and signal transduction in biological systems: mechanistic implications for ecological physical chemistry. In: Bonati L, Cosentino M, Lasagni M et al, eds. Trends in Ecological Physical Chemistry, Proceedings of the 2nd International Workshop on Ecological Physical Chemistry. Amsterdam: Elsevier, 1993:1-16.

Acknowledgments

This investigation was supported in part (PHJN) by the Dutch Foundation for Chemical Research (SON) with financial aid from the Dutch Organization for Scientific Research (NWO). We gratefully acknowledge Dr. A.H.J. Donné (Institute for Plasma Physics (FOM), Association Euratom-FOM) for his aid in designing the figures. We express our great gratitude to Dr. R.H. Davies (Welsh School of Pharmacy) for critically reading the manuscript and to Professor Dr. M. Freissmuth (Pharmacological Institute, University of Vienna), Dr. J.-W. de Gier (Arrhenius Institute, Stockholm University), Ing. H. Moereels (Janssen Research Foundation) and Dr. N.P. Shankley (James Black Foundation) for stimulating discussions.

"[..] Da ich ein Rationalist bin, so will ich niemanden bekehren. Ich will auch nicht den Namen der Freiheit dazu mißbrauchen, irgend jemanden zu einem Rationalisten zu machen. Aber ich möchte andere zum Widerspruch herausfordern; ich möchte, wenn möglich, andere dazu anregen, die Dinge in einem neuen Licht zu sehen, damit jeder in möglichst freier Meinungsbildung *seine* eigene Entscheidung treffen kann. Jeder Rationalist muß mit Kant sagen: Die Philosophie kann man nicht lehren—höchstens das Philosophieren; das heißt, die kritische Einstellung."

Sir Karl R. Popper

"Traveller, there isn't any road. On the way, one's path is hewn."

A. Machado

"[..] since no paradigm ever solves all the problems it defines and since no two paradigms leave all the same problems unsolved, paradigm debates always involve the question: Which problems is it more significant to have solved?"

Thomas S. Kuhn

PART I

Bioenergetical Mechanisms, Hydrogen-Bonded Chains and Proton Transfer Mechanisms

CHAPTER 1

Bioenergetics

1.1 INTRODUCTION

Bioenergetics: The Chemiosmotic Theory*

The narrow definition of bioenergetics comprises the mechanism by which the energy made available by the oxidation of substrates, or by the absorption of light, is coupled to 'uphill' reactions such as the synthesis of ATP from ADP and P_i. A broader definition also comprises transport processes. The largest part of ATP synthesis is associated with membrane-bound enzyme complexes found in the plasma membranes of prokaryotes, the inner membrane of mitochondria and the thylakoid membrane of chloroplasts. Despite the different natures of their primary energy sources, these membranes, so-called energy-transducing membranes, have a related evolutionary origin which makes them the core of bioenergetics.

Each energy-transducing membrane possesses two distinct types of proton pumps, a primary and a secondary pump. Depending on the energy source used by the membrane, respiratory or photosynthetic, primary pumps differ largely from each other while secondary pumps are highly conserved.

The *primary pumps* found in respiratory membranes use electron-transfer chains to catalyze the 'downhill' transfer of electrons from substrates to acceptors such as O_2. The energy associated with these processes is then used to generate a proton gradient, which is electrochemically denoted by $\Delta\mu_{H^+}$. In their turn, photosynthetic bacteria use light as an energy source to build up this

---

* *An excellent introduction to the field of bioenergetics can be found in ref. 1. Additional review material is available from refs. 2-5, as well as from refs.* loc. cit.

proton gradient, while chloroplasts not only produce a gradient from absorbed light but also drive electrons 'uphill' from water to acceptors such as $NADP^+$. The side of the membrane to which the protons are pumped by the primary pump is denoted the P (positive) side the opposite side is referred to as the N (negative) side. A bioenergetic convention is to convert $\Delta\mu_{H^+}$ into units of electronic potential (mV) and to refer to this as the protonmotive force, Δp.

In contrast to the great variety in mechanisms for generating Δp, the major consumer of this proton electrochemical gradient is the highly conserved *secondary proton pump* ATP synthase, also called (proton-translocating) ATPase. ATP synthases are found in mitochondria, chloroplasts, aerobic and photosynthetic bacteria, as well as in those bacteria that lack a functional respiratory chain and function on glycolysis. If this secondary pump is expressed in a membrane in the absence of a primary pump, it will hydrolyze ATP to ADP and P_i and thus pump protons in the same direction (uphill) as the primary pump. However, and this is the essence of bioenergetics, in case the primary pump generates a high enough Δp, it forces the secondary pump to reverse pumping and to actually synthesize ATP from ADP and P_i. Although we use bioenergetic conventions calling ATP synthase a secondary pump throughout this review, the word 'pump' could be misleading. In our opinion, general understanding would be improved if this enzyme is seen as a kind of turbine converting the energy of a proton gradient into another energy form, i.e., a high ATP/ADP ratio.

Both the primary and the secondary pumps are tightly coupled to proton translocation and work in a synergistic manner. The pumps are linked via a so-called proton circuit (Fig. 1.1). Figure 1.1 shows a hypothetical model for chemiosmotic coupling; only one primary pump, in this case bacteriorhodopsin (BR), is coupled via a proton circuit with the secondary pump, ATP synthase. Under influence of light, BR starts to pump protons against a gradient, thereby generating a Δp. In reality, Δp is not solely dependent on primary pump activity, but is influenced by many other factors, such as transport across membranes of substrates and/or ions (in particular ATP) and the permeability of the membrane. The secondary pump, $F_1 \cdot F_0$-ATP synthase, is seen to consume the Δp. The

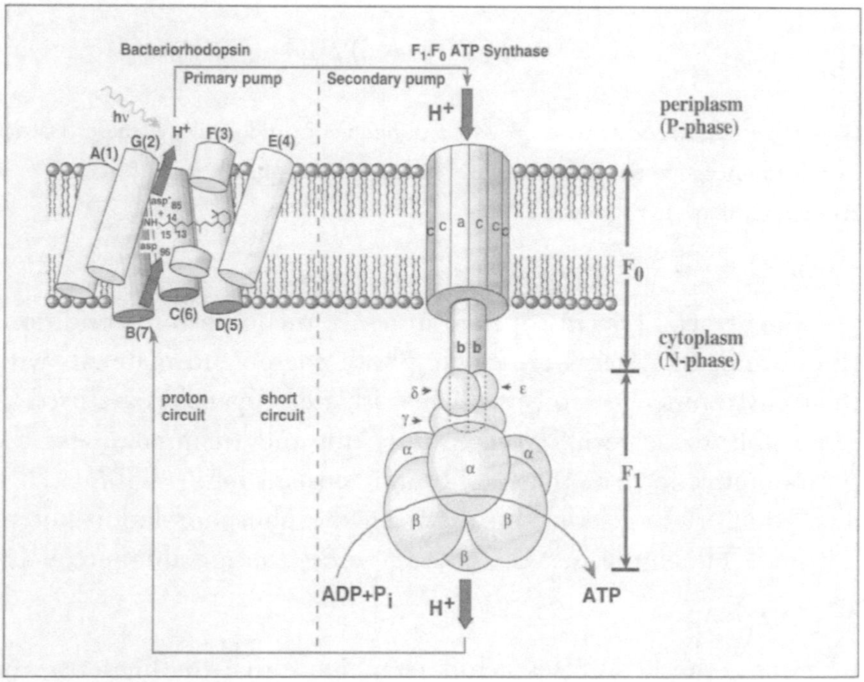

Fig. 1.1. Draft of a proton circuit: A primary pump, bacteriorhodopsin, pumps protons from the inside to the outside. A proton gradient is established resulting in an N-side (negative, inside) and a P-side (positive, outside). The secondary pump, ATP synthase, uses the proton gradient to synthesize ATP from ADP and P_i by translocating the protons via the transmembrane F_0 portion of the protein to the catalytic F_1 site. Also, bacteriorhodopsin translocates protons, most likely via hydrogen-bonded chains observed in the three-dimensional structure.[6,7] The proton circuit can be short-circuited by uncouplers, such as CCCP or FCCP. (Reprinted with permission from: Nederkoorn PHJ, Timmerman H, Donné-Op den Kelder GM. Does the ternary complex act as a secondary proton pump and a GTP synthase? Trends Pharmacol Sci 1995; 16:156-161.)

ATP synthase contains a membrane-spanning proton channel F_0, which translocates the protons delivered by the primary pump via its protein interior to the soluble part F_1 (provided a high enough Δp). The F_1 unit is directly connected to F_0 and uses the protons for ATP synthesis at its catalytic site.

The Protonmotive Force Δp

It is important to understand the nature of the aforementioned protonmotive force Δp. This proton electrochemical gradient consists of two components: one due to the difference in concentration of protons across the membrane (ΔpH) and the other due to the electrical potential built up over the membrane ($\Delta\psi$):

$$\Delta p = -(\Delta\mu_{H^+}) / F = +\Delta\psi - 2.3\cdot(RT / F)\cdot\Delta pH \quad (1.1)$$

(F is the Faraday constant; ΔpH is the difference in pH; and Δψ the difference in membrane potential between the P-phase and N-phase).

At 25°C, Eq. 1.1 becomes:

$$\Delta p\ (mV) = \Delta\psi - 59\Delta pH \quad (1.2)$$

Bioenergetic systems operate under non-equilibrium conditions (they are open systems exchanging both energy and materials with their environment). However, when isolated organelles are used, it is possible to achieve a true equilibrium and to interconvert the protonmotive force to the equilibrium constant for μ(ATP)/μ(ADP). The equilibrium relationship between the phosphorylation potential for ATP synthesis, ΔG_p, and Δp across the membrane equals:

$$\Delta G_p = n\cdot F\cdot\Delta p \quad (1.3)$$

where n is the H^+/ATP stoichiometry. Note that the higher n, the higher ΔG_p (for more details see refs. 9,10). Eqs. 1.1-1.3 are straightforward. However, in vivo one is not dealing only with non-equilibrium conditions. Furthermore, considerable controversy has arisen concerning the interpretation of the calculated stoichiometric factors (see ref. 1, and refs. *loc. cit.* for further details).

Finally, there are two classes of protons. First, there are the so-called 'scalar' or 'chemical' protons which are involved in the chemical reaction under consideration at physiological pH:

$$ADP^{3-} + P_i^{2-} + H^+ \rightleftharpoons ATP^{4-} + H_2O \quad (1.4)$$

(H^+ is a 'chemical' proton; P_i^{2-} is HPO_4^{2-}; it is noted that Mg^{2+} (not shown in Eq. 1.4) has a stabilizing influence on the formation of ATP)

Secondly, there are translocated or so-called 'vectorial' protons, which traverse the membrane and generate a low pH. The distinction between these different classes will be used as such in subsequent sections. Note that in Figure 1.1 only vectorial protons are drawn, although chemical protons are involved in the process as well and influence the protonmotive force Δp.

1.2 PROTON SEMICONDUCTORS AND ENERGY TRANSDUCTION: HYDROGEN-BONDED CHAINS

Morowitz[11] was the first to link ATP synthesis with proton transport via so-called hydrogen-bonded chains (HBCs, Fig. 1.2) consisting of proton-translocating functionalities such as protein

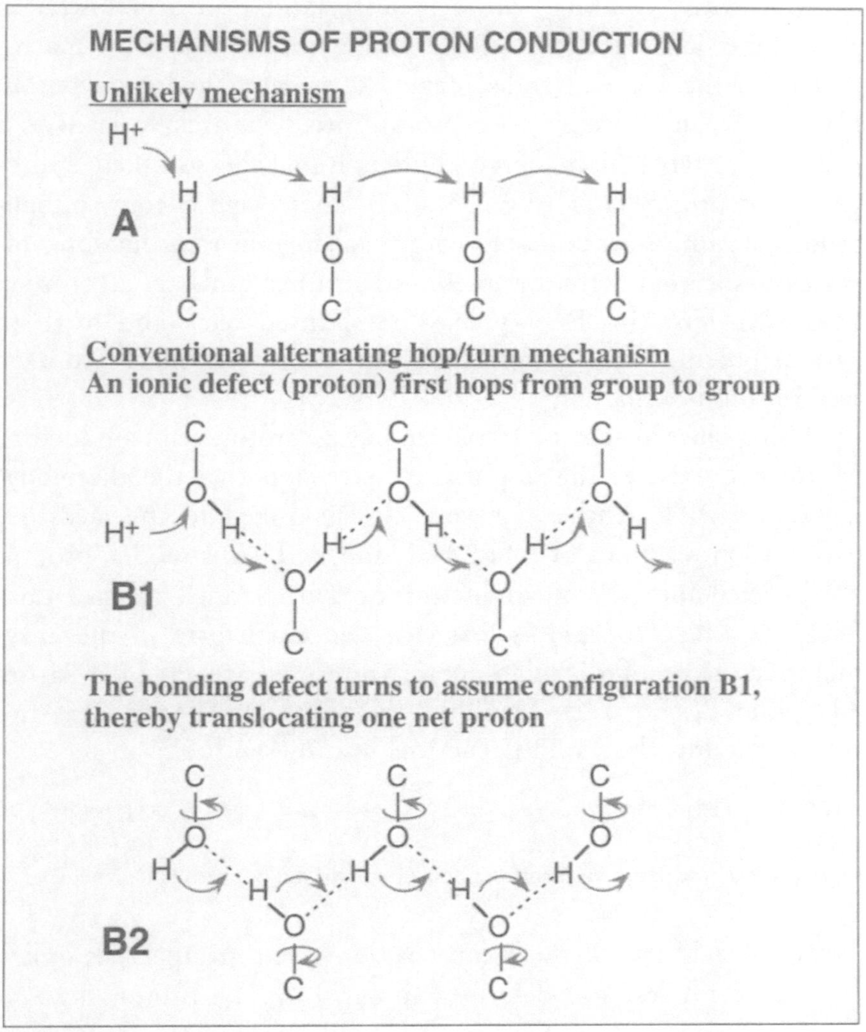

Fig. 1.2. Proton conduction mechanisms.[12] A: Unlikely mechanism for translocating a proton via a hydrogen-bonded chain (an HBC). B: Conventional mechanism, i.e., an alternating hop/turn mechanism, thereby translocating one proton: B_1: Ionic defect hops *from group to group (arrows denote movement of protons, not electrons); B_2: The bonding defect turns to retain conformation B_1. (Reprinted with permission from: Nederkoorn PHJ, Timmerman H, Donné-Op den Kelder GM. Does the ternary complex act as a secondary proton pump and a GTP synthase? Trends Pharmacol Sci 1995; 16:156-161.)*

amine and hydroxyl groups. A model is proposed in which a gated proton semiconductor (HBC) present within the protein ATP synthase carries protons from a high to a low proton electrochemical potential and allows them to interact specifically with well-defined substrate molecules such as ADP in order to synthesize ATP.

Morowitz[11] combines protochemistry and electrochemistry. In general his ideas are: (i) Within and in between certain macromolecules there are extended chains of ordered hydrogen bonds, HBCs. (ii) The injection of a proton into such a chain leads to a reasonably stable high-energy configuration, the so-called hydrogen bond proton fault (Fig. 1.2). (iii) These high-energy configurations are universal units of energy exchange in transduction processes associated with organelle structures and act as *proton semiconductors*. (iv) Energy may be shunted back and forth to various potential storage forms through the intermediate hydrogen bond proton faults.

Using the concept of HBCs acting as proton semiconductors, Morowitz[11] reaches the important conclusion that the thermodynamics of ATP synthesis is very closely linked to the acid-base dissociation reactions of ATP, ADP and P_i. In this model, protons are injected into a semiconducting proton channel, reach a catalytic site where the pH is low (±3) and participate in the enzymatic synthesis of the high-energy intermediate H_2ATP^{2-} from ADP and P_i (Eq. 1.4). H_2ATP^{2-} then dissociates into $2H^+$ and ATP^{4-} driving the overall reaction toward synthesis:

$$ATP^{4-}_{(bound)} + 2H^+ \rightleftharpoons H_2ATP^{2-}_{(bound)} \longrightarrow ATP^{4-}_{(free)} + 2H^+ \quad (1.5)$$

(The protons are vectorial protons; bound stands for bound to a protein.)

Thus, the acid-base dissociation reaction given in Eq. 1.5 is linked to a chemical reaction described in Eq. 1.4. The proton flow allows the organelle to utilize coupled acid-base reactions in the same way as one can use a coupled oxidation-reduction reaction when a linked electron flow is present. So again, protochemistry parallels conventional electrochemistry.

Morowitz[11] summarizes Eqs. 1.1-1.5:

$$ADP + P_i + ESE \rightleftharpoons ATP \tag{1.6}$$

(ESE represents the electrochemically stored energy and consists of ΔpH and $\Delta\psi$ components; cf. Eqs. 1.1 and 1.2).

Morowitz's ideas explicitly require the presence of transmembrane proton semiconductors. All known protonic semiconductors exhibit extended networks of hydrogen bonds, also called proton wires. Proteins with a suitable fraction of hydrogen-bonding side chain functionalities (the hydroxyl group of serines, threonines and tyrosines, the thiol group of cysteines, the carboxyl group of aspartic and glutamic acids, the amide of asparagines and glutamines, the amine of lysines, the guanidine group of arginines and the tautomeric forms of histidines) may fold in the lipophilic environment of the membrane in such a way that chains of hydrogen bonds (HBCs) are formed that span the membrane and conduct protons across it. To traverse a membrane takes approximately 20-25 successive hydrogen bonds.[13] *A single break in the chain destroys its connectivity and conductivity*, because the barrier for proton transport is now dramatically enhanced.

The primary pump bacteriorhodopsin (Fig. 1.1), a membrane protein that functions as a light-driven proton pump in the purple membrane of *Halobacterium halobium*, is the first proton-translocating protein in which chains of hydrogen bonds have been identified. They appear to be present in the boundary region between the α-helices,[6,7] indicating that in bacteriorhodopsin the protons indeed are translocated via HBCs. Very recently, the crystal structure at 2.8 Å resolution of cytochrome *c* oxidase from *Paracoccus Denitrificans* was determined.[14] Also within this protein, which is the terminal enzyme of most respiratory chains, elucidation of parts of the proton pathway(s) has now been claimed. However, despite the structural information, there is still considerable discussion about the HBC(s) present in this enzyme. Iwata et al[14] assume the presence of two distinct proton pathways, whereas Williams[15] concludes that the presence of a second proton-pumping path is still very uncertain. Because protons hop at very short distances (smaller than 1 Å) and require rotations of various donor and acceptor groups (N.B. including water molecules, *vide infra*), the nature of the proton pathway(s) in a 2.8 Å crystal structure of only one conformational state of cytochrome *c* oxidase is still uncertain.[15] These kinds of crystal structures can, however, add to our understanding and trigger new investigations into the nature of how HBCs are constructed in proton-translocating enzymes.

An HBC could involve waters of hydration as well. Putative roles of water molecules in proton flux mechanisms in a number of systems are discussed by Deamer and Nichols (ref. 16, and refs. therein). In bacteriorhodopsin, the constitutive role of waters of hydration in forming HBCs has been demonstrated experimentally.[7] HBCs involving protein interiors need to be distinguished from liquid water pores. A protein HBC consists of only a single chain of hydrogen bonds, whereas a liquid water pore requires a much larger channel area.[13] Moreover, a liquid water pore is permeable to other than hydrogen ions and the pore would lose the semipermeability associated with an HBC, thereby short-circuiting the protonmotive force.[13]

An HBC conducts a proton via a conventional alternating hop/turn mechanism.[12] The concept of hydrogen-bonded chains acting as proton wires is based on conductivity and dielectric studies in hydrogen-bonded crystals, in particular ice. The standard theory of proton conduction in such crystals consists of a two step process.[17] One step involves the passage of an ionic defect (an excess proton *hops* from one polar group to another polar group) (Fig. 1.2 B_1). There are two kinds of ionic defects: an excess proton on a group (positive defect) or a proton deficiency on a group (negative defect). The HBC is then blocked to the passage of another ionic defect until a second process occurs. This second step is the conductance of a so-called bonding defect. Bonding defects, or Bjerrum faults, may be of two types: no protons at a bond (so-called L fault, or Leerstelle) or two protons at a bond (so-called D fault, or Doppelbesetzung). In this second step a group adjacent to the bonding defect *turns* (Fig. 1.2 B_2). Only after a turn is processed, can another proton be transported.

The hop/turn mechanism can begin with either a hop or a turn, but thereafter hops and turns must strictly alternate. Transport of the bonding defect is absolutely necessary for repeated proton transport along the same chain. This hop/turn mechanism also explains why in water the mobility of protons is much higher than that of any other ion (apart from hydroxyl anions, for which a similar mechanism holds).[18] Furthermore, an increase in the water temperature from 0°C to 100°C decreases the abnormally high mobility of H^+ (and OH^-), due to a distortion of the water structure.[18] The optimal tetrahedral arrangement of H-bonds around a central oxygen atom becomes disrupted when going from ice to fluid water and hinders an optimal proton transport via a hop/turn mechanism.

In general, proton pumping across a membrane by a *primary pump* such as bacteriorhodopsin (BR) (Fig. 1.1) consists of four separate charge movements via two hop/turns.[12] The first proton moves via an ionic defect (hop) along an HBC from the catalytic site to the periplasm followed by a bonding defect (turn); a second proton (an ionic defect) then moves from the cytoplasm along the HBC (hop) to the catalytic site again followed by a turn. For BR this means a deprotonation of the Schiff base formed between retinal and Lys^{216} (catalytic site) upon activation by light. The liberated proton moves as a hop towards Asp^{85} to be liberated at the periplasmic side. The HBC between Lys^{216} and Asp^{85} then turns. Subsequently, a proton from the cytoplasm is injected into the HBC connecting Asp^{96} and Lys^{216} (i.e., the catalytic Schiff base is reprotonated) whereupon a turn follows. A more detailed description of BR is given in the next chapter. These proton movements result in a proton gradient over the membrane with a negative inside (N-side, cytoplasm) and a positive outside (P-side, periplasm). For extensive descriptions concerning possible mechanisms of coupling proton wires to active sites, the reader is referred to Nagle et al.[12,13,19,20]

The transfer of a defect from solution to a hydrogen-bonded chain at normal pH is thought to cost at least 3 kcal.[20] For the thermodynamical and kinetic aspects concerning hydrogen-bonded chains, which can perform the function of kinetically competent proton wires in membrane bioenergetics, the reader is referred to Nagle et al[12,13,19,20] and to Onsager.[17] For example, it is demonstrated that the energy difference between the states B_1 and B_2 in Figure 1.2 can be up to 12 kcal for an asymmetric chain consisting of 20 H-bonds (i.e., a chain in which OH---OH---OH is lower in energy than HO---HO---HO or vice versa), which corresponds to membrane potentials as high as 500 mV for protons.[13] Furthermore, an asymmetric HBC can speed up the transport of the rate limiting defect.[12] Before discussing the function of proton conduction in primary and secondary proton pumps, the proton circuit itself will be addressed.

1.3 THE PROTON CIRCUIT

In Figure 1.1 a proton circuit has been drafted in which the primary pump bacteriorhodopsin is coupled to the secondary pump ATP synthase. Recently, experimental evidence has shown that a proton conducting circuit can indeed be compared to a classical electron conducting circuit.[21] The authors gathered evidence that protons pumped across membranes by a primary pump such as bacteriorhodopsin do not readily diffuse into the aqueous bulk phase (which would correspond to the delocalized theory), but move much faster along the membrane surface (Fig. 1.3, localized theory). The data suggest that protons released by a source (a primary pump) can efficiently diffuse along the membrane surface over rather large distances towards a sink (secondary pump) without dissipation losses into the aqueous bulk.

Heberle et al[21] observed that a pH sensor positioned at the membrane surface at an average distance of 240 nm from the proton releasing site, bacteriorhodopsin, detected the released protons eight times faster than the pH probe in the bulk water phase at an average distance of only 17 nm from the proton-ejecting source (Fig. 1.3). As a consequence, proton sources and sinks both embedded in the lipid bilayer phase do not need to be in close spatial contact in order to form a proton circuit. Heberle et al[21] suggest that the origin of the fast diffusion of protons along the surface might be the interplay of protonation and deprotonation of amino acids and lipid head groups. Also, proton transfers via water molecules restrained at the membrane surface are assumed to be involved. So, implicitly, the authors[21] state that a well-ordered HBC (Fig. 1.2) is formed along the membrane, which conducts protons via an alternating hop/turn mechanism. The conduction along the membrane is thought to be much faster than in the aqueous bulk, because at physiological temperature the hydrogen-bonded chains are partially broken in the bulk, resulting in a lower proton conductance.[18]

An important implication of the above study is that proton conduction via a localized circuit will result in only negligible proton changes in the aqueous bulk phase. Only when the proton

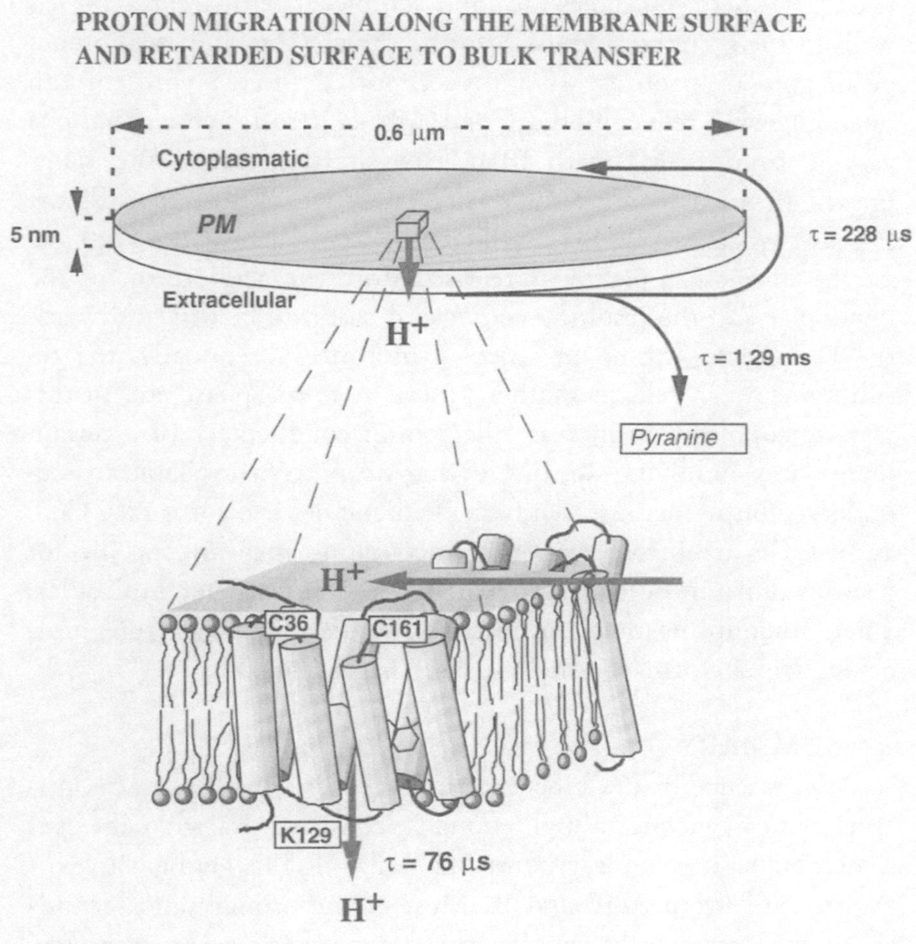

Fig. 1.3. Schematic draft of the purple membrane containing the integral protein bacteriorhodopsin and attachment sites of the pH-indicating dye fluorescein (boxed); adapted from Heberle et al (1994). Pyranine detects pH changes in the bulk water phase. Time constants τ were: 76 µs for proton release to the extracellular membrane surface, 228 µs for lateral proton transfer and 1.29 ms for surface to bulk transfer. N.B. within the experimental settings chosen, protons need to cross a membrane rim, which is shown to be the rate limiting step in translocation.[21] In vivo, such rims are absent and lateral diffusion will therefore be faster than 228 µs. (Reprinted with permission from: Heberle J, Riesle J, Thiedemann G et al. Proton migration along the membrane surface and retarded surface to bulk transfer. Nature 1994; 370:379-382.)

translocating ability of the secondary pump is blocked or impaired, while in the meantime a primary pump is active, will protons diffuse into the medium and affect the bulk pH. As long as a circuit can be established, no large pH differences are to be expected.

The proton circuit can be short-circuited by uncouplers, also called protonophores, such as carbonyl cyanide *m*-chlorophenylhydrazone (CCCP) or carbonyl cyanide *p*-trifluoromethoxyphenylhydrazone (FCCP). These molecules are lipid soluble in both their acidic as well as their conjugated base form, thereby making membranes permeable to protons. If a protonmotive force is present, an uncoupler will reduce both $\Delta\psi$ and ΔpH to zero (at concentrations varying from 1 nM up to 10 μM) by shuttling across the membrane.[1] In other words, $CCCP^-$ or $FCCP^-$ will diffuse towards the P-phase, whereas CCCPH and FCCPH accumulate at the N-phase. At the N-phase a proton is released from the acidic form of the uncoupler and the resulting conjugated base will be driven towards the P-phase by the Δp to accept a proton. Subsequently, the resulting acid can release another proton at the N-phase etc. In this way, uncouplers can increase the proton conductance of a certain membrane, nullifying Δp and leaving no protons available to secondary pumps. Because both the acidic and basic form of CCCP and FCCP are involved in this uncoupling, they are specific for protons and can be considered highly specific carrier molecules. Their uncoupling mechanism must be distinguished from molecules which form channels inside the membrane.

1.4 SUMMARY

The essence of the bioenergetical theory is that, in case a primary pump generates a high enough electrochemical gradient over a membrane, a secondary pump can consume this gradient to synthesize ATP from ADP and P_i. Most of the primary and secondary pumps are tightly coupled to proton translocation (the electrochemical gradient is then called the protonmotive force), thereby forming a so-called proton circuit. Experimental evidence was obtained[21] that such a proton circuit is closed, i.e., the pumped protons are efficiently translocated over the membrane surface towards a sink (or secondary pump) without dissipation losses into the aqueous bulk. The proton circuit can thus be compared to a classical electron-conducting circuit. Protons are conducted via hydrogen-bonded chains. These chains act as proton wires, have a polar character and display a two-step transport mechanism (hop/turn). Also, water molecules can be part of hydrogen-bonded chains, which are found inside proteins as well as along the membrane surface. Wherever water molecules are found in a relatively fixed tetrahedral arrangement (as is the case at the membrane surface), proton transport via the hop/turn mechanism is optimal. Uncouplers or protonophores, such as CCCP or FCCP, can reduce Δp to zero by short-circuiting the proton circuit.

References

1. Nicholls DG, Ferguson SJ. In: Bioenergetics 2. London:Academic Press, 1992.
2. Boyer PD. A perspective of the binding change mechanism for ATP synthesis. FASEB J 1989; 3:2164-2178.
3. Pedersen PL, Amzel LM. ATP synthases. Structure, reaction center, mechanism, and regulation of nature's most unique machines. J Biol Chem 1993; 268:9937-9940.
4. Penefsky HS, Cross RL. Structure and mechanism of F_0F_1-type ATP synthases and ATPases. Adv Enzymol 1991; 64:173-214.
5. Tonomura Y. F_1-ATPase. In: Energy-transducing ATPases—Structure and kinetics. Avon: Cambridge University Press, 1986:141-183.
6. Henderson R, Baldwin JM, Ceska TA et al. Model for the structure of bacteriorhodopsin based on high-resolution electron cryomicroscopy. J Mol Biol 1990; 213:899-929.
7. Dencher NA, Büldt G, Heberle J et al. Light-triggered opening and closing of a hydrophobic gate controls vectorial proton transfer across bacteriorhodopsin. NATO ASI Ser, Ser B 1992; 291:171-185.
8. Nederkoorn PHJ, Timmerman H, Donné-Op den Kelder GM. Does the ternary complex act as a secondary proton pump and a GTP synthase? Trends Pharmacol Sci 1995; 16:156-161.
9. Krab K, van Wezel J. Improved derivation of phosphate potentials at different temperatures. Biochim Biophys Acta 1992; 1098:172-176.
10. Slater EC, Rosing J, Mol A. The phosphorylation potential generated by respiring mitochondria. Biochim Biophys Acta 1973; 292:534-553.
11. Morowitz HJ. Proton semiconductors and energy transduction in biological systems. Am J Physiol 1978; 235:R99-R114.
12. Nagle JF, Tristram-Nagle S. Hydrogen bonded chain mechanisms for proton conduction and proton pumping. J Membrane Biol 1983; 74:1-14.

13. Nagle JF, Morowitz HJ. Molecular mechanisms for proton transport in membranes. Proc Natl Acad Sci USA 1978; 75:298-302.
14. Iwata S, Ostermeier C, Ludwig B et al. H. Structure at 2.8 Å resolution of cytochrome *c* oxidase from *Paracoccus Denitrificans*. Nature 1995; 376:660-669.
15. Williams RJP. Purpose of proton pathways. Nature 1995; 376:643.
16. Deamer DW, Nichols JW. Proton flux mechanisms in model and biological membranes. J Membrane Biol 1989; 107:91-103.
17. Onsager L. In: Whalley E, Jones SJ, Gold LW, eds. Physics and Chemistry of Ice. Ottawa: Royal Society, 1973; 7-12.
18. Robinson RA, Stokes RH. In: Electrolyte solutions. 2nd ed. London: Butterworths, 1970.
19. Nagle JF, Mille M. Molecular models of proton pumps. J Chem Phys 1981; 74:1367-137.
20. Nagle JF, Mille M, Morowitz HJ. Theory of hydrogen bonded chains in bioenergetics. J Phys Chem 1980; 72:3959-3971.
21. Heberle J, Riesle J, Thiedemann G et al. Proton migation along the membrane surface and retarded surface to bulk transfer. Nature 1994; 370:379-382.

CHAPTER 2

Primary and Secondary Proton Pumps

2.1 PRIMARY PUMPS—A SPECIAL CASE: BACTERIORHODOPSIN

Primary pumps differ highly depending upon the energy source used by the membrane. For example, the respiratory chains of mammalian mitochondria act as oxidation-reduction driven proton pumps, which transfer electrons from the $NAD^+/NADH$ couple to the O_2/H_2O couple. The respiratory chain consists of more than 20 discrete carriers of electrons which are mainly grouped into four polypeptide complexes: NADH-ubiquinone oxidoreductase, succinate dehydrogenase, ubiquinol-cytochrome *c* oxido-reductase and cytochrome *c* oxidase. Three of these complexes are involved in proton translocation.

Photosynthetic organisms capture light in order to drive the primary proton pump. Bacteriorhodopsin (BR) (cf. Fig. 1.1) belongs to this class of light-driven pumps and deserves our special attention since (i) Timms et al[1,2] strongly suggest that the majority of G protein-coupled receptors (GPCRs) are capable of promoting a proton transfer mechanism just like bacteriorhodopsin (BR); (ii) the mutation-induced activated state of the β_2 adrenergic receptor[3] (see Part II) can be explained with a similar mechanism as proposed for the signal transduction by BR; and (iii) the three-dimensional structure of BR contains a seven membrane-spanning α-helix motif and shares this motif with the class of GPCRs; in the literature BR is therefore frequently used as a template molecule for generating three-dimensional models of GPCRs (ref. 4, and refs. *loc. cit.*).

Since H^+ translocation through bacteriorhodopsin is not associated with electron transfer, this protein differs from other light-driven or respiratory proton pumps. Henderson et al[5] suggest that pK_a changes in the Schiff base formed between the chromophore retinal and the ε-amino group of Lys^{216} must act as the means by which light energy is converted into proton pumping. The proton channel is formed by 26 residues originating from five helices.[5] From mutation studies (ref. 5 and refs. *loc. cit.*) it has been inferred that a minimal model for proton pumping involves two aspartic acids (Asp^{85} and Asp^{96}) and the Schiff base in which Lys^{216} and retinal participate. Asp^{212} is functioning as the counterion of the Schiff base in the resting state, whereas during activation Asp^{85} and the Schiff base transiently get closer, leading to the formation of an H-bond[6] (Fig. 2.1). Asp^{85} lies closest to the periplasm (P-phase), Lys^{216} approximately in the middle of the membrane, while Asp^{96} is on the pathway from the cytoplasm (N-phase) to the Schiff base. In the resting state (retinal being in the all-*trans* state), the pK_a of the Schiff base (~10) is significantly higher than the pK_a of a similar model compound in an aqueous solution (~7). Also Asp^{85} and Asp^{96} are indicated to have abnormal pK_a values, i.e., Asp^{85} has an abnormally low pK_a value and is fully deprotonated in the resting state, whereas Asp^{96} has an abnormally high pK_a value and is therefore fully protonated despite the high cytosolic pH. These abnormal pK_a values are believed to be caused by local protein surroundings.[5] The reversible dissociation and association of a proton from and to the Schiff base is thought to drive the proton pump, as explained in the next paragraph.

Absorption of a photon causes a crucial change in the conformation of retinal from all-*trans* to 13-*cis*. This localized structural change enables a proton transfer from the fully protonated Schiff base to the fully deprotonated Asp^{85}. This induces further conformational changes in the protein at the expense of energy stored in retinal and prevents Asp^{85} from reprotonating the Schiff base. Subsequently, the excellent proton-accepting abilities of the Schiff base are restored by further conformational changes, whereas Asp^{85} regains its low pK_a value and releases its proton into the periplasm (P-phase). The initial (resting) state of bacteriorhodopsin is restored by reprotonation of the Schiff base with the aid of the fully protonated Asp^{96}; retinal relaxes to its all-*trans* conformation and Asp^{96} accepts a proton from the cytoplasm (N-phase). Approximately 80% of the charge movement related to the proton transport occurs during the last part of the photocycle, which is consistent with the observation that the lower proton channel, connecting the Schiff base to the cytoplasm, is narrow and hydrophobic in nature (principal electrical barrier for H^+ movement) and the upper extracellular part is wider and more hydrophilic (lower barrier).[5] In conclusion, the light-induced movement of the Schiff base linkage of the chromophore to Lys^{216} and/or alterations in the tertiary structure of BR should be considered as the driving force for the translocation of vectorial protons and the generation of a protonmotive force. As mentioned earlier, the complete process consists of two separate hop/turn alternations (cf. chapter 1).

Dencher et al[6] report direct evidence for tightly bound water molecules close to the chromophore binding site and suggest that they might participate in the aforementioned process of light-induced vectorial proton transport. Moreover, they indicate that these water molecules are part of a network of hydrogen bonds (N.B. HBC) via which the translocated protons are transported from the cytoplasm (N-phase) via Asp^{96}, Lys^{216} and Asp^{85} to the periplasm (P-phase). In the resting state, bacteriorhodopsin is unable to translocate protons. This might be due to an interruption of the conducting pathway in the resting state over a distance of about 7Å by the presence of hydrophobic amino acids (Phe^{219}, Leu^{43}; and Val^{49}; Fig. 2.1), which form a so-called hydrophobic gate closed under resting conditions and prevent proton conductance via the hydrophilic HBC.[6] Obviously, it is essential that the proton pathway is closed as long as BR is inactive. This property necessarily prevents uncontrolled proton backflow in the resting state of the protein and an undesired collapse of any electrochemical proton gradient present across the membrane. Only during the short period of conformational changes induced by the absorption of light must the hydrophobic gate open and possibly be filled with a transient population of 2-3 water molecules, allowing transportation of protons (Fig. 2.1).

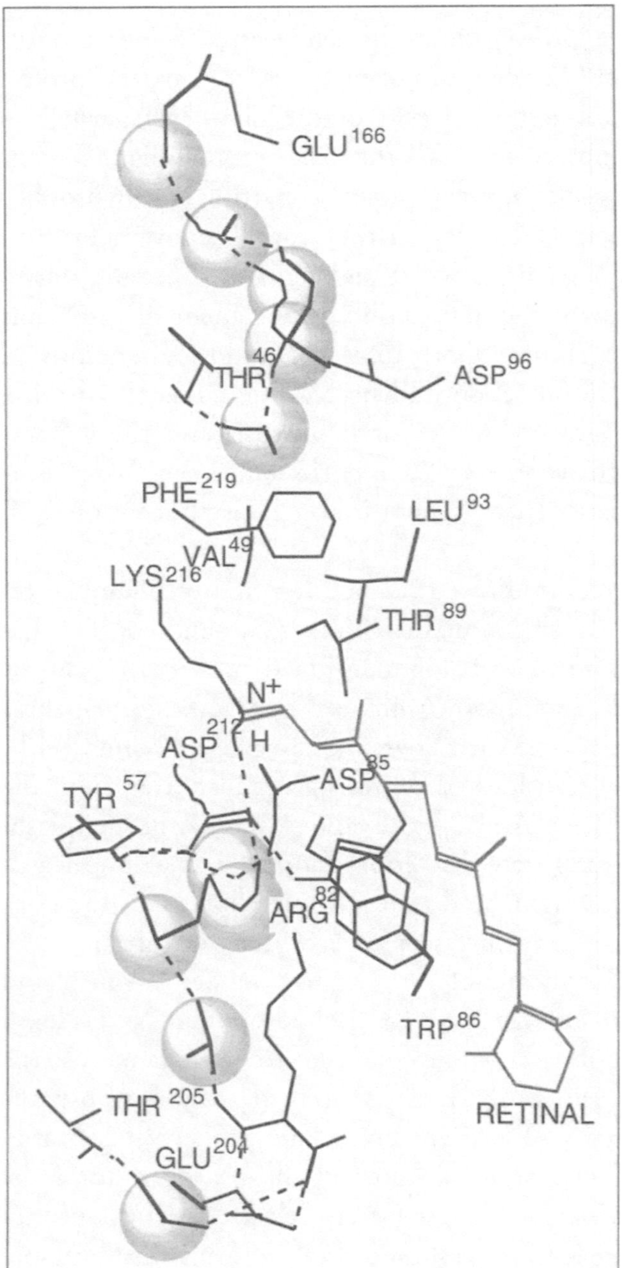

Fig. 2.1. Hydrogen-bonded chain in BR. Water molecules (shaded balls) are seen to take part in a proton wire. A hydrophobic gate (Phe[219], Leu[93] and Val[49]) prevents the consumption of the proton-motive force. (Reprinted with permission from: Dencher NA, Büldt G, Heberle J et al. Light-triggered opening and closing of a hydrophobic gate controls vectorial proton transfer across bacteriorhodopsin. NATO ASI Ser, Ser B 1992; 291:171-185.)

2.2 SECONDARY PUMPS, ATP SYNTHASES

The secondary pump ATP synthase is the major consumer of the protonmotive force, Δp, generated in a respiratory or photosynthetic system. Of special importance are the structural features of the separate F_1 and F_0 units and the role of proton conduction in the ATP synthesis reaction. The next chapter deals with the chemistry of ATP synthesis and hydrolysis reactions.

The proton translocating ATP synthases are generally known as $F_1 \cdot F_0$-ATPases (also called F-type ATPases or ATP synthases) and differ from P-type (or E_1-E_2) ATPases such as the (Na^+/K^+)-translocating ATPases, which are ubiquitous in higher eukaryotes and have a catalytic cycle that involves a covalent attachment of the γ-phosphate group to the protein. $F_1 \cdot F_0$-ATPases also differ from V (vacuolar) ATPases, which can pump protons across internal membranes such as synaptic vesicles in neurones. It should be noticed that the bioenergetics of neurotransmitter release and uptake can also be interpreted within the concept of ion fluxes.[7,8] For a complete classification of ATPases the reader is referred to Pedersen and Amzel.[9]

The essential function of ATP synthases is to utilize the protonmotive force, Δp, generated by a primary pump in order to maintain the mass action ratio ATP/ADP 7 to 10 orders of magnitude away from equilibrium. Only in the case of fermentative bacteria is ATP utilized for transport purposes. Thus, in all cases except the last one, the function of the $F_1 \cdot F_0$-complex is to synthesize ATP rather than to hydrolyze it. However, sometimes the enzyme is referred to as the proton-translocating ATP*ase* since in the absence of a Δp it hydrolyzes ATP. Thus, the intact $F_1 \cdot F_0$-ATP synthase constitutes an enzyme complex that functions as an H^+-translocator using the protonmotive force generated by a primary pump in order to synthesize ATP. The membrane-bound F_0 unit takes care of the necessary proton transfer, the F_1 unit (which can display ATPase activity when decoupled from F_0) is responsible for nucleotide binding and subsequent enzymatic activity (ATP synthesis or hydrolysis).

Structural Features of F_1

The F_1 unit contains five different types of polypeptides, known as α, β, γ, δ, ε (cf. Fig. 1.1). The α- and β-chains are arranged alternately in an assembly possessing 3-fold symmetry, consistent with an $\alpha_3\beta_3\gamma\delta\varepsilon$ stoichiometry. Furthermore, F_1 from mitochondria contains a sixth subunit that inhibits both ATPase activity and ATP synthesis but dissociates from F_1 during purification procedures.[10,11] The soluble F_1 unit, which is connected to the membrane-bound F_0 portion, is responsible for the knobs seen under an electron microscope. In all energy-conserving membranes ATP is hydrolyzed or synthesized on the side of the membrane from which the knobs project (i.e., the N-phase). Thus F_1 faces either the mitochondrial matrix, the bacterial cytoplasm or the chloroplast stroma.

F_1 units contain six nucleotide binding sites identified by X-ray data of F_1-ATPase from bovine heart mitochondria.[12] Only three sites readily exchange bound nucleotide for medium nucleotide and have a catalytic role (ref. 12, and refs. therein). Three non-catalytic sites fail to readily release bound nucleotide during MgATP hydrolysis or synthesis[13] and evidence for a control function is given by Xue and Boyer.[14] The non-catalytic sites lie at the interface between α- and β-subunits, whereas the catalytic sites are predominantly present in β-subunits with some contributions from side chains in the α-subunits.[12] In the presence of labelled nucleotides, catalytic sites are rapidly labelled and the non-catalytic sites slowly.[15] Binding of ATP at non-catalytic sites markedly promotes hydrolysis of ATP at the catalytic sites. Chemical modification of the β-chain (in highly conserved sequences) inactivates the enzyme, while the β-chains from some sources may have ATPase activity in the absence of other chains. Summarizing these data, we conclude that the β-subunits are the main sites for catalytic adenine nucleotide binding and subsequent ATP synthesis or hydrolysis.

Both the α- and β-chains possess sequences indicative of adenine nucleotide binding.[16] Sequence conservation for the other three chain types is less pronounced and their role is unknown, except that they are all needed for the reconstitution of an $F_1 \cdot F_0$-ATP synthase able to translocate protons. The F_1 sequence possesses the conserved nucleotide-binding motif, GxxxxGKT/S (one for each nucleotide binding site).[12] Also X-ray structures of other nucleotide-binding proteins such as $p21^{ras}$, adenylate kinase, RecA, elongation factor Tu, the α-unit of the G protein (G_α) and several others, reveal the same motif (ref. 12, and refs. therein). The crystal structure of F_1 furthermore explains why the nucleotide-binding sites on the β-subunits are catalytically active in contrast to the non-catalytic sites at the interface of the α- and β-subunits. The catalytic sites appear to possess a Glu pointing to the terminal phosphate, whereas the non-catalytic sites have a Gln pointing away from the phosphate. This Glu probably activates a water molecule, enabling a nucleophilic attack on the γ-phosphate.

Structural Features of F_0

The membrane-embedded F_0 unit is hydrophobic. All three F_0 chains (a, b and c) are required in the in situ $ab_2c_{10\text{-}12}$ stoichiometry in order to create a membrane-traversing proton channel. The sequence of the a-chain suggests the presence of 6-8 alpha helices. Although the b-chain has a hydrophobic N-terminus, the remainder of its sequence is hydrophilic, which is probably necessary to form a spindle onto which the F_1 complex is mounted. The c-chain is mainly hydrophobic but contains one aspartic acid, located in the middle of the bilayer. Mutation of this Asp to an Asn results in an enzyme which is unable to conduct protons.[16] It is tempting to propose the presence of an hydrogen-bonded chain with a crucial role for this aspartic acid.

The Function of Proton Translocation

The key question is of course how the translocation of protons drives ATP synthesis. Nicholls and Ferguson[16] describe the framework for a model in which protons enter F_0 from the P-phase, pass down the electrical field created by the membrane potential and accumulate in the "reaction chamber" located on F_1, where ATP is synthesized. Thus, regardless of whether the major component of Δp is a $\Delta\psi$ (mitochondria) or a ΔpH (thylakoids), a large proton gradient will be established between the reaction chamber and the N-phase (cf. Fig. 1.1). Both Nicholls and Ferguson[16] as well as Boyer[15] emphasize that due to the high H^+ concentration at the bottom of F_0, protons could bind to an accessible site on F_1 even if it has a relatively low pK_a. Conversely, the proton will be able to dissociate from the binding site after a proposed conformational change which makes the site accessible to the N-phase (so-called indirect mechanism, *vide infra*). This deprotonation occurs even if during the process the proton becomes much more tightly bound (i.e., the pK_a of the site increases) since the pH of the N-phase may be 3-4 pH units higher than at the bottom of the F_0 unit.

During the conversion from a loosely to a tightly bound proton (a change from a low to a high pK_a value) Gibbs energy can be transferred to the protein, allowing it to adopt a high energy conformation. In this way, the protonmotive force generated by the primary pump might induce a 'high-energy state' in the ATP synthase, possibly accompanied by changes in the affinity for adenine nucleotides. As experimental evidence is available that tightly bound ATP is formed from tightly bound ADP and P_i with little input from Gibbs energy[15] (see next chapter for further details), the Δp-induced 'high-energy state' of the complex might be necessary to lower the affinity of the catalytic binding site for ATP and allow the nucleotide to leave the complex. This mechanism, favored by Boyer and co-workers, is called the *indirect* pathway.[15] Also, Nicholls and Ferguson[16] conclude that, unlike bacteriorhodopsin's reaction mechanism which involves a proton flux through the catalytic site (Schiff base of retinal), the balance of evidence is against a *direct* involvement of the translocated protons in the ATP synthesis reaction itself (see next chapter and refs. *loc. cit.* for further details). However, Mitchell[17,18] prefers the *direct* mechanism in which protonation of the phosphate oxygen drives ATP synthesis. Morowitz[19] also prefers a *direct* mechanism in which the release of two translocated protons drives the reaction towards synthesis (Eq. 1.5).

2.3 SUMMARY

Bacteriorhodopsin, a well-known primary proton pump, contains a membrane-spanning seven α-helical motif. Inside this protein, a hydrogen-bonded chain is found, consisting of polar amino acids as well as of water molecules. In the resting state, this hydrogen-bonded chain is interrupted by a so-called hydrophobic gate, which inhibits proton conductance. The hydrophobic gate thus prevents the collapse of any protonmotive force. Absorption of a photon causes the retinal molecule, bound to bacteriorhodopsin as a Schiff base, to change from all-*trans* to 13-*cis*, resulting in localized conformational changes inside bacteriorhodopsin. These very short-lived localized structural changes induce subsequent alterations in pK_a values of specific amino acids (including the Schiff base between a lysine residue and retinal), the opening of the hydrophobic gate (the gate becomes transiently filled with water molecules) and the translocation of protons. Per photon, one proton is pumped across the membrane, thereby generating an electrochemical proton gradient.

An intact $F_1 \cdot F_0$-ATP synthase, a secondary pump, can use this protonmotive force to synthesize ATP. There are two theories available to explain how translocation of protons can drive ATP synthesis: (i) via the direct mechanism in which protonation of phosphate oxygens drives ATP synthesis, (ii) via the indirect mechanism in which translocated protons induce conformational changes inside $F_1 \cdot F_0$-ATP synthase, causing a 'high-energy state' necessary for tightly bound ATP to leave the protein.

References

1. Timms D, Wilkinson AJ, Kelly DR et al. Interactions of Tyr^{377} in a ligand-activation model of signal transmission through β_1-adrenoceptor α-helices. Int J Quant Chem: Quant Biol Symp 1992; 19:197-215.
2. Timms D, Wilkinson AJ, Kelly DR et al. Ligand-activated transmembrane proton transfer in β_1-adrenergic and m_2-muscarinergic receptors. Receptors and Channels 1994; 2:107-119.
3. Samama P, Cotecchia S, Costa T et al. A mutation-induced activated state of the β_2-adrenergic receptor. Extending the ternary complex model. J Biol Chem 1993; 268:4625-4636.
4. Hoflack J, Trumpp-Kallmeyer S, Hibert M. Re-evaluation of bacteriorhodopsin as a model for G protein-coupled receptors. Trends Pharmacol Sci 1994; 15:7-9.
5. Henderson R, Baldwin JM, Ceska TA et al. Model for the structure of bacteriorhodopsin based on high-resolution electron cryomicroscopy. J Mol Biol 1990; 213:899-929.
6. Dencher NA, Büldt G, Heberle J et al. Light-triggered opening and closing of a hydrophobic gate controls vectorial proton transfer across bacteriorhodopsin. NATO ASI Ser, Ser B 1992; 291:171-185.
7. McMahon HT, Nicholls DG. The bioenergetics of neurotransmitter release. Biochim Biophys Acta 1991; 1059:243-264.

8. Njus D, Kelly PM, Harnadek GJ. Bioenergetics of secretory vesicles. Biochim Biophys Acta 1986; 853:237-265.
9. Pedersen PL, Amzel LM. ATP synthases. Structure, reaction center, mechanism, and regulation of nature's most unique machines. J Biol Chem 1993; 268:9937-9940.
10. Pedersen PL, Schwerzmann K, Cintron N. Regulation of the synthesis and hydrolysis of ATP in biological systems: Role of peptide inhibitors of proton-ATPases. Curr Top Bioenerg 1981; 11:149-199.
11. Tonomura Y. F_1-ATPase. In: Energy-transducing ATPases—structure and kinetics. Avon: Cambridge University Press, 1986:141-183.
12. Abrahams JP, Leslie AGW, Lutter R et al. Structure at 2.8 Å resolution of F_1-ATPase from bovine heart mitochondria. Nature 1994; 370:621-628.
13. Penefsky HS, Cross RL. Structure and mechanism of F_0F_1-type ATP synthases and ATPases. Adv Enzymol 1991; 64:173-214.
14. Xue Z, Boyer PD. Modulation of the GTPase activity of the chloroplast F_1-ATPase by ATP binding at noncatalytic sites. Eur J Biochem 1989; 179:677-681.
15. Boyer PD. A perspective of the binding change mechanism for ATP synthesis. FASEB J 1989; 3:2164-2178.
16. Nicholls DG, Ferguson SJ. In: Bioenergetics 2. London:Academic Press, 1992.
17. Mitchell P. A chemiosmotic molecular mechanism for proton-translocating adenosine triphosphatases. FEBS Lett 1974; 43:189-194.
18. Mitchell P. Biochemical mechanism of protonmotivated phosphorylation in $F_0 \cdot F_1$ adenosine triphosphate molecules. In: Lee CP, Schatze G, Dallner G, eds. Mitochondria and Microsomes. Reading: Addison Wesley, 1981:427-457.
19. Morowitz HJ. Proton semiconductors and energy transduction in biological systems. Am J Physiol 1978; 235:R99-R114.

CHAPTER 3

ATP Hydrolysis and Synthesis Mechanisms

3.1 THE CHEMISTRY OF ATP HYDROLYSIS AND SYNTHESIS

A way to extend our knowledge of the mechanism of ATP synthesis catalyzed by the $F_1{\cdot}F_0$-complex is to elucidate the mechanism of its reverse reaction, the ATPase hydrolysis reaction. As we have indicated, ATP hydrolysis activity of the F_1-unit can be observed when the $F_1{\cdot}F_0$-complex is decoupled from a protonmotive force or when F_1 is decoupled from F_0.

A Classical S_N2 Reaction

The ATP hydrolysis reaction mechanism can either be studied by labelling the solvent ($H_2{}^{18}O$) (schematically shown in Eq. 3.1a) and adding cold ATP, or by labelling the ligand, (^{18}O)ATP, and working in cold solvent (cf. Eq. 3.1b).

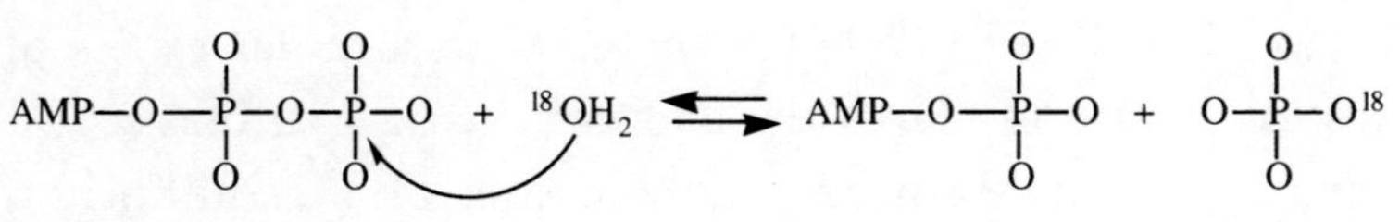

(3.1a)

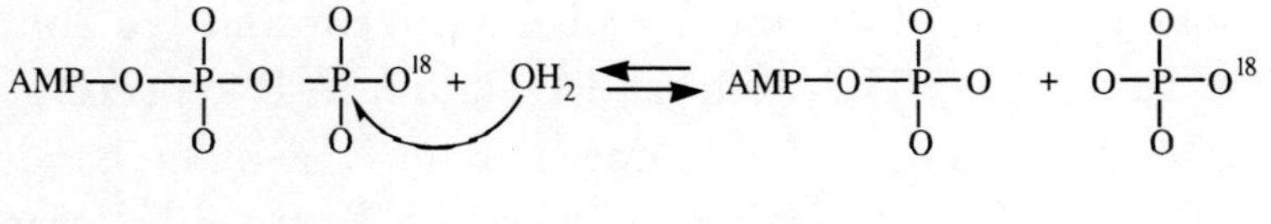

(3.1b)

^{18}O from labelled water is built into (γ-)P_i rather than into ADP and the reaction probably consists of a chiral inversion at the γ-P_i. The hydrolysis reaction could therefore be similar to a classical S_N2 reaction involving cleavage of the bond between the bridging oxygen and the γ-phosphorus. One oxygen on the leaving P_i then stems from water (Eqs. 3.1a-b). Further evidence for this inversion reaction was obtained using the ATP-analogue adenosine 5'-(3-thiotriphosphate) stereospecifically labelled with ^{18}O in the γ-position.[1] The ^{18}O exchange experiments also provide evidence that the hydrolysis activity of the F_1-unit is to some small extent reversible. The reverse (synthesis) reaction even occurs in the absence of a Δp, indicating that the protonmotive force is not an absolute prerequisite to condense ADP and P_i to ATP (*vide infra*).[2-7]

ATP/ADP Exchange Reactions and Cooperation

Nucleotide exchange reactions have been observed at various ATP concentrations. Even at very low concentrations ATP is bound at the catalytic sites of the F_1-unit with apparent dissociation constants of approximately 10^{-10} M.[2] At substrate concentrations far below those required for half-maximum velocity, a catalytic site on F_1 is already fully occupied. Furthermore, addition of ATP appears to promote the release of tightly bound ADP from bacterial F_1-ATPase, suggesting that ATP binding at one catalytic site promotes ADP release at another.[3] Also, myosin ATPase binds ATP tightly (10^{-11} M), but is unable to catalyze net ATP synthesis since it cannot release the tightly bound ATP.[4] These findings are in agreement with observations on ATPases from chloroplasts, which can synthesize tightly bound ATP from tightly bound ADP when exposed to a high concentration of medium P_i.[5]

Penefsky and Cross[6] report that when a small amount of labelled ATP (ATP-γ-^{32}P < 10^{-10} M) is mixed with a molar excess of F_1, ATP is rapidly bound (K_d 10^{-12}) and enzymatic hydrolysis proceeds slowly. At the high affinity catalytic sites of F_1, equilibrium concentrations of ATP, ADP and P_i will be present and the sum of bound reactants (i.e., ATP, ADP and P_i bound to F_1) is equivalent to the total amount of ATP added. If, on the other hand, excess ATP is added so that alternate sites on F_1 are filled, the release of bound ADP and P_i is dramatically accelerated.[6] From these findings co-operativity between the catalytic sites is apparent. The catalytic sites co-operate negatively with respect to nucleotide binding, but positively regarding ATP hydrolysis. Reasoning *a contrario*, this co-operativity explains detectable ATP *synthesis* at *low* ATP concentrations.

THE GIBBS FREE ENERGY FOR ATP SYNTHESIS AND RELEASE

Since the equilibrium constant for the *non-enzymatic* ATP hydrolysis reaction** at pH 7 and 10^{-2} M [Mg^{2+}] equals 10^5 M, the reverse *non-enzymatic* (ATP synthesis) reaction is expected to be undetectable.[7] In contrast, *enzymatic* catalysis in the presence of F_1 enhances ATP synthesis significantly. It therefore seems that enzymatic synthesis of ATP requires little or no additional energy input. Also experiments on "de-energized" submitochondrial particles (within these particles, Δp is absent) led to the conclusion that ATP is formed without any energy input.[4] The ΔG value for enzymatically catalyzed ATP synthesis is thus close to zero and the equilibrium constant close to unity (Eq. 3.2a). Nearly all the energy delivered by the protonmotive force can thus be used to release tightly bound ATP from the catalytic site via a conformational change in the protein, as discussed in the previous chapter (*indirect* mechanism). Δp decreases the affinity for ATP from a value of 10^{-12} M in the absence of a Δp to a sufficiently loose binding such that ATP can dissociate in the presence of a normal N-phase concentration of nucleotide (Eq. 3.2b). The decrease in binding affinity induced by Δp has been estimated to be 10^7 M, resulting in a dissociation constant of 10^{-5} M.[8]

$$ADP^{3-}_{(bound)} + P_{i\,(bound)} + H^+ \rightleftarrows ATP^{4-}_{(bound)} + H_2O \qquad (3.2a)$$

(with ΔG 0, K_{eq} 1 M^{-1})

$$ATP^{4-}_{(bound)} + \Delta p \longrightarrow ATP^{4-}_{(free)} \qquad (3.2b)$$

Besides the protonmotive force contributing to ATP release, ADP binding to adjacent catalytic sites is also known to facilitate ATP release.[6] Nicholls and Ferguson[7] note furthermore that a Na^+ electrochemical potential can also drive ATP synthesis, as is observed for the Na^+-translocating ATP synthases in the bacterium *Propionigenium modestum.* In general, it is accepted that the

** *$[\Sigma ADP] \cdot [\Sigma P_i] / [\Sigma ATP]$, where each concentration is the total sum of the various ionic species of each component including those complexed to Mg^{2+}.*

H^+/ATP (or Na^+/ATP) stoichiometry is approximately 3 (cf. Eq. 1.3).

3.2 DIRECT OR INDIRECT MECHANISM FOR ATP SYNTHESIS?

MUTUAL INTERACTIONS BETWEEN F_1 AND F_0

Experiments with the intact $F_1 \cdot F_0$ complex have revealed that both the antibiotic oligomycin and *N,N'*-dicyclohexylcarbodiimide (DCCD) block the F_0 proton channel but at the same time drastically affect the high-affinity binding of ATP to F_1. DCCD is known to modify a glutamate in the c-subunit of F_0 and to reduce ATP binding to F_1.[6] On the other hand, binding of trinitrophenyl-ADP at catalytic sites of the F_1-unit of the chloroplast ATP synthase appears to open the pathway for proton translocation through the F_0-unit.[9] Only conformational changes relayed from the F_0-unit to the F_1-unit and vice versa seem to be able to explain these findings.[4,6,7]

These experiments indicate that energy-requiring conformational changes initiated in F_0 can be transmitted through intervening subunit structures to catalytic sites on F_1, causing an enhanced release of tightly bound ATP.[4] As early as 1978, Morowitz proposed that protons injected into an ordered HBC could cause faults which are associated with structural rearrangements.[10] We therefore propose that the chemical modification due to DCCD obstructs, or at least alters, an HBC which runs from F_0 into the reaction chamber on F_1, inducing chain modifications and a decreased affinity for ATP.

T-, L- AND O-SITES

In general, it is accepted that there are three catalytic sites (probably one on each β subunit), which can exist in an open (O), loose (L), or tight (T) conformation (refs. 7,11, and refs. *loc. cit.*). A model has been presented (Fig. 3.1) in which proton-induced conformational changes cause a T-site with bound ATP to become an O-site and release ATP (Eq. 3.2b), while at the same time an L-site with loosely bound ADP and P_i changes into a T-site where the substrates are tightly bound, allowing bound ATP to be formed (Eq. 3.2a). It is suggested that each of these catalytic sites has at any instant a different conformation, but that they all pass sequentially through the same conformations. Boyer (ref. 4, and refs. therein) was the first to suggest a rotational movement of the catalytic subunits relative to F_0. For the possible existence of such a rotation, the reader is also referred to Abrahams et al.[11]

The Direct and Indirect Mechanisms Integrated

ATP synthesis has been proposed to occur via either a *direct* or an *indirect* pathway (see previous chapter). The sharp dualism

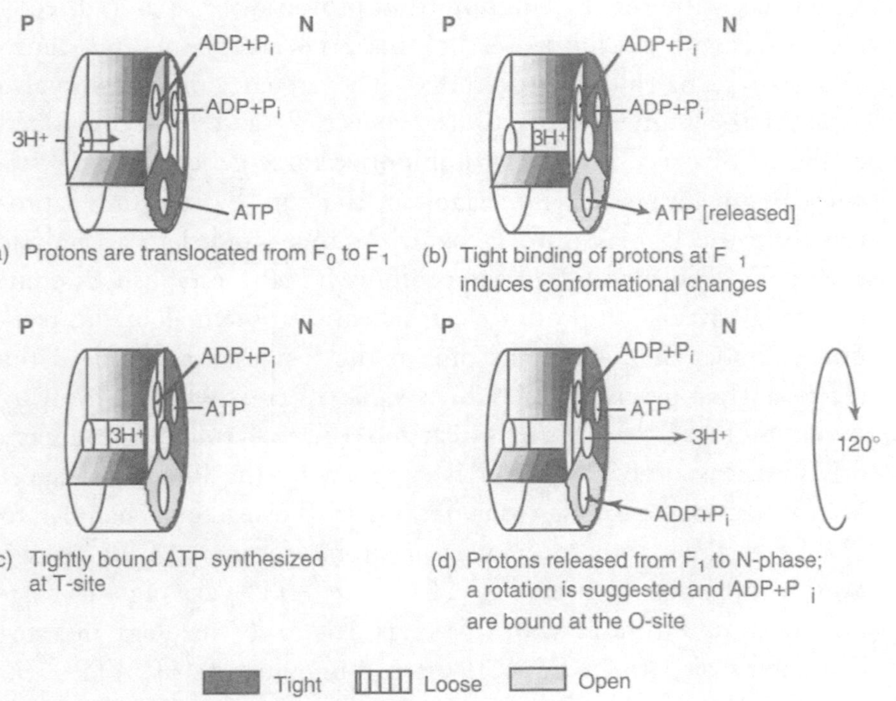

Fig. 3.1. Three catalytic subunits of the F_1 portion of F_0·F_1-ATP synthase participating in the synthesis of ATP from ADP and P_i. The three sites can exist in an open (O), loose (L), or tight (T) conformation (ref. 7, and refs. loc. cit.). Translocated protons interact with the F_1 portion causing structural changes, thereby forcing the T-site with bound ATP to become an O-site and to release ATP, forcing the O-site to become an L-site with loosely bound ADP and P_i, and at the same time forcing the L-site to change into a T-site where substrates are tightly bound and ATP is formed (Eq. 3.2). It is suggested that each of these catalytic sites has at any instant a different conformation and that they all pass sequentially through the same conformations. In this respect a 120° rotation–depicted in (d)–is necessary to regain the initial situation (a). P and N denote the P-phase and N-phase, respectively. (Reprinted with permission from: Nicholls DG, Ferguson SJ. In: Bioenergetics 2. London: Academic Press, 1992.)

between these two mechanisms can be overcome when they are integrated into one new mechanism by considering Morowitz's HBCs for proton transfers[10] in connection with a closed (localized) proton circuit.[12]

In its most simple form, a hydrogen-bonded chain acts as a passive proton wire conducting protons efficiently and quickly to a suitable site for chemical action. Morowitz[10] used a simplified form of this concept for coupling ATP synthesis to proton transport. In our attempt to integrate the direct and indirect mechanism, we derived a model in which a proton wire is located inside the F_0 subunit of ATP synthase and ends in a reaction chamber associated with the F_1 subunit; this proton wire is not directly connected with the N-phase (i.e., bacterial cytoplasm, mitochondrial matrix, or chloroplast stroma). The presence of this wire also explains the mutual interactions between F_1 and F_0 as described earlier in this section. The protonmotive force generated by a primary pump induces an ionic fault in the wire by injecting a proton from the P-phase into F_0, which is then guided to a catalytic site on F_1. The function of the proton wire is to establish an equilibrium between the proton electrochemical potential in the reaction chamber on F_1 and the one in the P-phase. The pH of the reaction chamber is lowered to a value of 3-4, which Morowitz shows to favor the synthesis reaction (Eq. 1.4). The free energy driving the synthesis reaction is supplied by the linked transport of two vectorial protons from pH 3-4 in the reaction chamber to pH 7-8 in the cytoplasm, matrix or stroma (chapter 1). These two protons are used to form H_2ATP^{2-} from ATP^{4-} driving the reaction towards synthesis (Eq. 1.5). Eqs. 1.4 to 1.5 explain the stoichiometry of 3 (n in Eq. 1.3) often found for the H^+/ATP ratio at physiological pH. One proton is of chemical origin (Eq. 1.4) and two are vectorial protons (Eq. 1.5). Deviation from the n value of 3 is to be expected since both Eqs. 1.4 and 1.5 are influenced by variations in pH and ADP/ATP ratio.

Integration of the direct and indirect mechanism is complete when we assume that the 'high energy' intermediate, H_2ATP^{2-} (Fig. 3.2a), attracts cytosolic water molecules from the N-phase, which enter the reaction chamber and induce conformational changes (Fig. 3.2b): the T-site becomes an O-site, the L-site

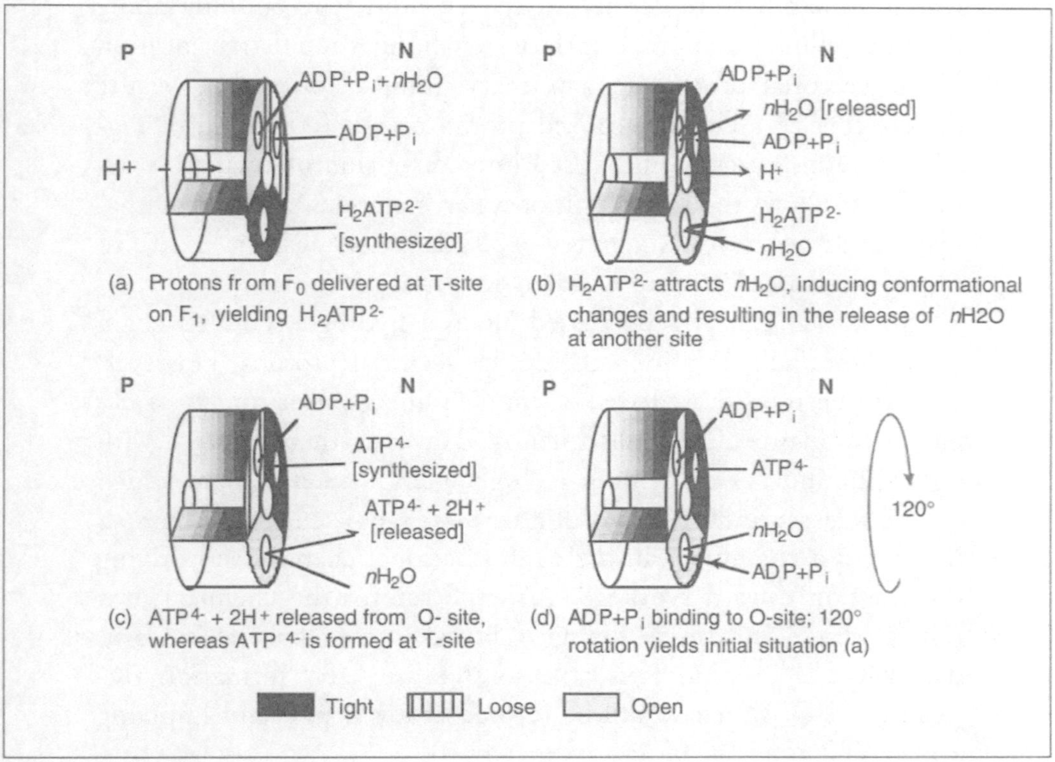

Fig. 3.2. A revised model for the involvement of three catalytic subunits of the F_1 portion of $F_0 \cdot F_1$-ATP synthase participating in the synthesis of ATP from ADP and P_i. Protons delivered by the proton motive force, Δp, interact with ATP^{4-} at the T-site to form the high-energy intermediate H_2ATP^{2-} (a) which becomes solvated (b). This solvation accounts for the conversion of the T-site into an O-site, while the L-site, binding ADP and P_i, becomes a T-site, and the O-site changes into an L-site with simultaneous release of n *H_2O molecules. The high energy intermediate H_2ATP^{2-} dissociates into $2H^+$ and ATP^{4-} (c). This process is entropy driven and accounts for the overall liberation of (free) ATP^{4-}, i.e., ATP^{4-} is synthesized. (picture adapted from Fig. 3.1). The mechanism ends with the alleged 120° rotation[4,11] (d) to regain the initial position (a). N.B. The waters entering the reaction chamber are connected to a proton wire running along the intracellular membrane surface[12] enabling a rapid translocation of the vectorial protons to a primary pump (the concept of a closed localized proton circuit (cf. Fig. 1.3). P and N denote P-phase and N-phase.*

becomes a T-site, and the O-site changes into the L-site, with simultaneous release of *n* water molecules). These structural alterations open the reaction chamber (pH 3-4) to the N-phase (pH 7-8) and the two vectorial protons can diffuse along the membrane surface (Fig. 1.3) in the direction of a primary pump, thereby driving the reaction towards ATP synthesis. Thus, the high energy intermediate becomes solvated in order to liberate $2H^+$ and ATP^{4-} (large entropy gain), which accounts for the release of bound ATP (Fig. 3.2c). We assume that the water molecules entering the reaction chamber are connected to a proton wire running along the intracellular membrane surface[12] enabling a rapid translocation of the vectorial protons to a primary pump. Notice that we use the concept of a closed localized proton circuit (cf. Fig. 1.3). Figure 3.2 summarizes the proposed interconversion of catalytic sites; the O-site allows the interaction of water molecules, and the mechanism ends with the suggested 120° rotation[4,11] (Fig. 3.2d) to regain the initial position drawn in Figure 3.2a.

Since Na^+ can also be passed along a hydrogen-bonded chain as an ionic fault, this ion can replace vectorial protons. However, Na^+ will be translocated via the unlikely mechanism for H^+ as depicted in Figure 1.2a. Subsequently, Na^+ can then interact with tightly bound ATP, yielding a 'high energy' intermediate which can be released into the cytosol after water molecules have entered the reaction chamber, causing conformational changes and driving the reaction toward synthesis. Also the role of the chemical proton (cf. Eq. 3.2a) can in our view be taken over by Na^+, since in the synthesis reaction it is able to stabilize OH^- formation (instead of H_2O, which would be formed at lower pH, and implying a Na^+/ATP ratio, n, of approximately 3).

3.3 ROTATIONAL MOVEMENT OF CATALYTIC SITES ON F_1

The aforementioned interconversion of the three catalytic sites on F_1, i.e., the T-site into an O-site, the L-site into a T-site and the O-site into an L-site (Figs. 3.1 and 3.2), has been brought into connection with a possible rotational movement of the F_1 unit with respect to F_0.[4,11] This is most easily illustrated for a secondary pump which forms a closed proton circuit with only one primary pump. In that case, proton translocation along the membrane surface from the secondary toward the primary pump (and vice versa) will always occur along a straight line connecting the two proton pumps. Proton release from the F_1 unit only occurs from the catalytic site which is in the open configuration, the O-site. Since all catalytic sites pass sequentially through the O-, L- and T-configurations, the orientation of the O-site with respect to the primary pump would continuously change in the absence of a rotational movement. Therefore, in order to establish an optimal orientation of the O-site with respect to the primary pump, a rotational movement of F_1 relative to F_0 is not unlikely. Possibly, the electrical field generated by the primary pump might be helpful in establishing the presumed rotation in such a way that the released vectorial protons from the O-site always pass along the same HBC towards a particular primary pump. Also the translocation of protons through the ATP synthase itself (in the direction of the tightly bound nucleotide) could contribute to a rotational movement.[4,11]

It is now important to mention that changes in an electrical field have been suggested to be the origin of a rotational movement of an α-helix present within the S4 segment of voltage-gated channels (e.g., for sodium-channels; see ref. 13). Depolarization of the membrane is thought to reduce the forces holding positive charges in their inward position in which they are paired with negative charges present on other transmembrane segments. The resulting rotational movement involves a screw-like motion of one of the S4 α-helices. Moreover, charge transfers are associated with this kind of motion. Thus, changes in electrical fields (such as generated by primary pump and secondary pump activity) as well as charge transfers/redistributions inside proteins could very well be the driving force of the rotational movements.

As the X-ray structure of F_1 is now known,[11] theoretical calculations might substantially contribute to the unraveling of the most likely mechanism (direct, indirect, integrated, with or without rotational movements).

E_s and E_h Conformations of the $F_1 \cdot F_0$-ATP Synthase

We end this résumé on proton translocating mechanisms inside $F_1 \cdot F_0$-ATP synthase with a speculative suggestion from Boyer[4] concerning the possibility that the F_1 unit exists in two conformational forms (Fig. 3.3). These states, indicated as E_s (s for synthesis) and E_h (h for hydrolysis), may be related to the exposure of groups involved in the proton translocation mechanism by F_0 (speaking in terms of HBCs, this exposure might correspond to an OH--OH--OH and HO--HO--HO state, respectively). Boyer's suggestion stems from observations that the capacity for ATP synthesis and ATP hydrolysis differ under different experimental conditions. ADP+P_i binding seems to favor the E_s (synthesis) form, whereas ATP binding prefers the E_h (hydrolysis) form. The conformational differences between the ATP synthase and ATPase form of the F_1 unit seem to be related to different conformations of the F_0 unit.

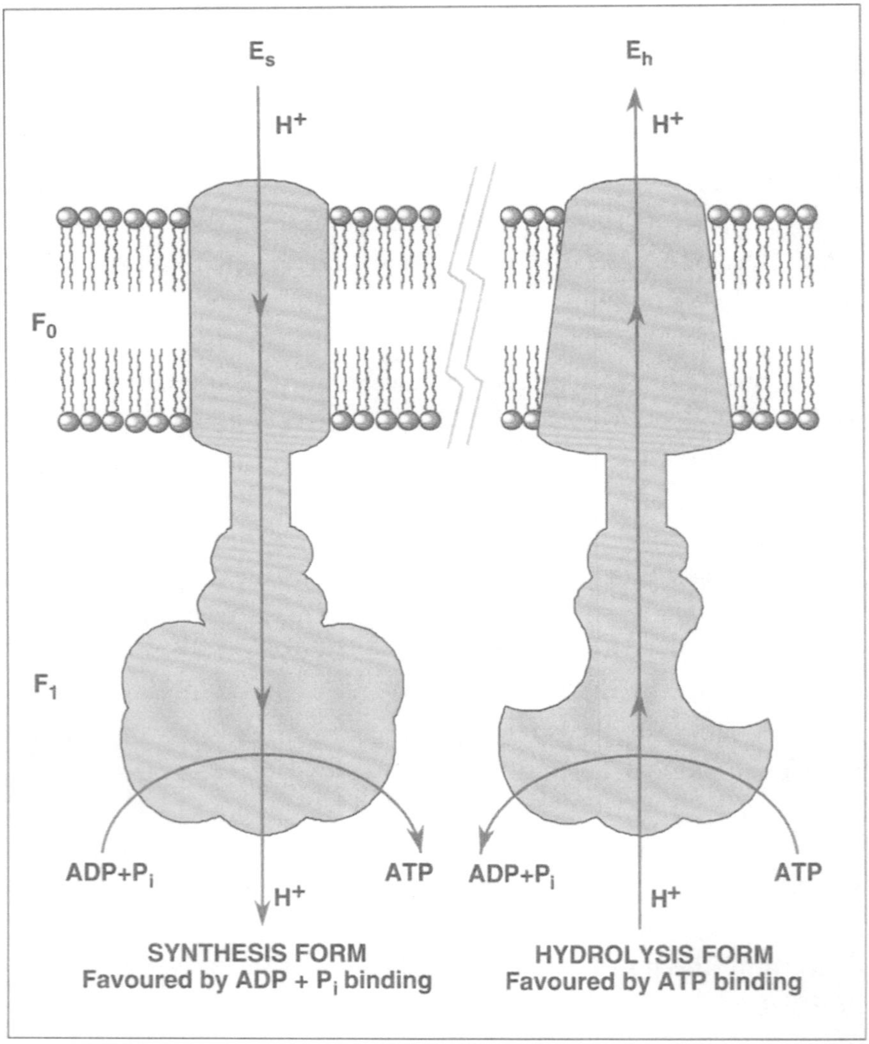

Fig. 3.3. Two forms of ATP synthase, E_s and E_h, as suggested by Boyer.[4] The E_s (synthesis) form is favored by ADP + P_i binding; the hydrolysis form E_h is favored by ATP binding. The orientation of the hydrogen-bonding functionalities in F_0 which are involved in the proton translocation process could possibly be different in the two enzyme forms and could therefore be involved in stabilizing E_s or E_h; e.g., the two states could correspond to an OH--OH--OH and an HO--HO--HO state, respectively. Note that the mechanisms depicted in both Figures 3.1 and 3.2 correspond with E_s activity.

The ATP synthesis mechanism with the interconverting L-, O- and T-sites (Figs. 3.1 and 3.2) only occurs when the protein is in the E_s form. E_s binds ADP and P_i tightly and is ready to couple ATP synthesis to proton translocation. E_s cannot hydrolyze ATP. In contrast, the E_h form can only hydrolyze ATP and drives proton translocation against a (weak) gradient. When the enzyme binds ADP without P_i in the presence of Mg^{2+}, a stabilization of the E_h form and thus inhibition of ATPase activity results.

3.4 SUMMARY

If tightly bound, ADP and P_i can be condensed to ATP almost without any energy input. The release of tightly bound ATP, however, requires energy which can be provided by the protonmotive force. The question was raised whether translocated protons cause conformational changes inside $F_1{\cdot}F_0$-ATP synthase, causing tightly bound ATP to be released (so-called indirect mechanism), or whether protons are transported to the active site, yielding the 'high-energy intermediate' H_2ATP^{2-} (so-called direct mechanism). This intermediate then becomes solvated and dissociates into $2H^+$ and ATP, accompanied by a large entropy gain. We offer an alternative in which both the direct and the indirect mechanism are integrated into one model.

In Part I, we have reviewed the field of bioenergetics to such an extent that the principles which we consider as important for an introduction to Part II are outlined. In this second part, proton translocation processes are brought into connection with a tentative model for G protein activation and signal transduction via GPCRs.[14] In Part I, we focused both on the molecular mechanisms underlying ATP synthesis and hydrolysis and on proton translocation processes occurring in primary and secondary proton pumps and along membrane surfaces. Whether protons are directly and/or indirectly involved in the ATP synthesis reaction, it is clear that proton translocation processes, energetically fed by the protonmotive force, are involved in the ATP synthesis.

Nicholls and Ferguson[7] end their *Bioenergetics 2* with the message that chemiosmotic energy transduction principles and methods for characterizing electron transfers and proton translocations do not only apply only to the bioenergetics of mitochondria, bacteria and chloroplasts. Indeed, we propose in Part II that these principles also seem to be applicable to signal transduction in GPCRs. Hence, ATP synthase might not be nature's most unique machine interconverting electrochemical and chemical energy,[15] because some of the bioenergetic principles seem also to be valid for the molecular mechanisms involved in signal transfer processes of GPCRs.

References

1. Reinstein J, Brune M, Wittinghofer A. Mutations in the nucleotide binding loop of adenylate kinase of *Escherichia coli*. Biochemistry 1988; 27:4712-4720.
2. Slater EC, Kemp A, Van der Kraan I et al. The ATP- and ADP-binding sites in mitochondrial coupling factor F_1 and their possible role in oxidative phosphorylation. FEBS Lett 1979; 103:7-11.
3. Adolfsen R, Moudrianakis EN. Binding of adenosine nucleotides to the purified 13S coupling factor of bacterial oxidative phosphorylation. Arch Biochem Biophys 1976; 172:425-433.
4. Boyer PD. A perspective of the binding change mechanism for ATP synthesis. FASEB J 1989; 3:2164-2178.
5. Feldman RI, Sigman DS. The synthesis of enzyme-bound ATP by soluble chloroplast coupling factor 1. J Biol Chem 1982; 257:1676-1683.
6. Penefsky HS, Cross RL. Structure and mechanism of F_0F_1-type ATP synthases and ATPases. Adv Enzymol 1991; 64:173-214.
7. Nicholls DG, Ferguson SJ. In: Bioenergetics 2. London: Academic Press, 1992.
8. Pérez JA, Ferguson SJ. Kinetics of oxidative phosphorylation in *Paracoccus denitrificans*. 1. Mechanism of ATP synthesis at the active site(s) of F_0F_1-ATPase. Biochemistry 1990; 29:10503-10518.
9. Wagner R, Ponse G, Strotmann H. Binding of 2'(3')-*O*-(2,4,6-trinitrophenyl)-adenosine 5'-diphosphate opens the pathway for protons through chloroplast ATPase complex. Eur J Biochem 1986; 161:205-209.
10. Morowitz HJ. Proton semiconductors and energy transduction in biological systems. Am J Physiol 1978; 235:R99-R114.
11. Abrahams JP, Leslie AGW, Lutter R et al. Structure at 2.8 Å resolution of F_1-ATPase from bovine heart mitochondria. Nature 1994; 370:621-628.
12. Heberle J, Riesle J, Thiedemann G et al. Proton migation along the membrane surface and retarded surface to bulk transfer. Nature 1994; 370:379-382.

13. Catterall WA. Voltage-dependent gating of sodium channels: correlating structure and function. Trends NeuroSci 1986; 9:7-10.
14. Nederkoorn PHJ, Timmerman H, Donné-Op den Kelder GM. Does the ternary complex act as a secondary proton pump and a GTP synthase? Trends Pharmacol Sci 1995; 16:156-161.
15. Pedersen PL, Amzel LM. ATP synthases. Structure, reaction center, mechanism, and regulation of nature's most unique machines. J Biol Chem 1993; 268:9937-9940.

PART II

Proton-Transferring Hydrogen-Bonded Chains and the Ternary Complex: The Ternary Complex as a GTP Synthase

CHAPTER 4

G Protein-Coupled Receptors and G Proteins

4.1 CHARACTERISTIC AND STRUCTURAL FEATURES OF G PROTEIN-COUPLED RECEPTORS

G protein-coupled receptors (GPCRs) belong to a large class of membrane-spanning receptors. GPCRs function as antennae for external signals consisting of chemical ligands or photons (in the case of opsins). Once the external signal is received, it is transduced to a cytosolic G protein. Subsequently, the G protein itself becomes activated. Hundreds of GPCRs have been sequenced, and they have been classified into three superfamilies:[1] (i) rhodopsin-like; (ii) secretin-like; and (iii) metabotropic glutamate receptors. Rhodopsin-like receptors can be characterized by conserved residue patterns: in TM1 (i.e., the first transmembrane α-helical domain), Gly-Asn; in TM2, Leu-x-x-x-Asp; at the end of TM3, Asp-Arg-Tyr; and in TM7, Asn-Pro. An important group of the rhodopsin-like family consists of the G protein-coupled amine receptors (α and β adrenoceptors, dopamine, histamine, muscarinic and serotonin receptors). Examples of other groups in this family are: peptide receptors (e.g., angiotensin II, bradykinin, chemotactic peptides, endothelins, neuropeptides, and tachykinins), olfactory receptors and opsins. Sequence conservation patterns in secretin-like and metabotropic glutamate receptors have not yet been identified.

In the absence of structural data (other than the amino acid sequences) on G protein-coupled receptors, hydrophobicity profiles together with information from biophysical, chemical, and site-directed mutagenic studies were used to reveal important structural features. The information led to a model in which GPCRs were suggested to be membrane-bound proteins which share a seven α-helix motif (reviewed in refs. 2,3). Further support for this idea was obtained from the three-dimensional structure of bacteriorhodopsin,[4] which is a functional homologue of the GPCR rhodopsin (in the sense that both transmembrane proteins bind retinal and are activated by light). The membrane-traversing portion of bacteriorhodopsin (BR) appears to consist of seven almost antiparallel α-helices. In 1993 firm experimental evidence was obtained that the GPCR rhodopsin indeed contains a seven α-helix motif.[5]

An analysis of GPCR sequences combined with information from projection maps on rhodopsin and bacteriorhodopsin reveals that the general arrangement of helices in these two proteins is similar. According to Baldwin[6] and Unger and Schertler,[7] the overall structure probably differs. However, Hoflack et al[8] assign the differences in the projection data for BR and rhodopsin to different orientations of these proteins in the two-dimensional crystals. The latter argument has been used to validate the numerous models which have been built for GPCRs using the three-dimensional structure of bacteriorhodopsin as a structural template (for a review see refs. 3,9).

The binding of a ligand to a GPCR is known to occur in between the transmembrane and sometimes extracellular domains, whereas the coupling of G proteins to the GPCR seems to be mediated by four regions of the receptor, including a C-terminal portion of the second intracellular loop, N- and C-terminal portions of the third intracellular loop and a part of the C-terminal tail (e.g., refs. 3,9-13 and refs. *loc. cit.*).

4.2 PROTON TRANSFERS IN GPCRs

Since the 1970s, receptor activation has been brought into connection with possible proton transfers. Ganellin[14] studied the relative populations of N^{τ}/N^{π} tautomers for the imidazole moiety of histamine and analogues, and he suggested that when the difference in free energy between the two tautomers would appear to be small, a proton transfer might be involved in histamine H_2 receptor activation. In line with this suggestion, Weinstein et al[15,16] developed a theoretical model in which binding of histamine to the H_2 receptor results in a neutralization of its side chain, followed by a double proton transfer from and to the imidazole group of histamine, resulting in a tautomeric shift of this ring.

Osman et al[17] observed that in the presence of 5-hydroxytryptamine (5-HT, serotonin) the proton-accepting ability of an imidazole ring in a histidine residue at the receptor changes. Based upon this finding an activation model for a 5-HT receptor was proposed in which protonation of a histidine residue is induced by binding of serotonin, which in turn is considered to trigger G protein activation.

Timms et al[18,19] developed a theoretical model for the activation of a series of amine receptors including β_1 adrenergic, m_2 muscarinic, and H_2 histaminergic receptors. The authors suggest that these receptors, and moreover the majority of GPCRs, are capable of catalyzing proton transfer mechanisms through the protein interior, which could be an integral part of the G protein activation mechanism. In their view, a receptor can act as a proton pump during excitation. The proton pumping ends with the delivery of at least one proton per agonist, intracellularly.

For the study of the activated histamine H_2 receptor Timms et al[19] assumed the dicationic species of histamine to be the active ionic species. However, it has been established that histamine binds as monocation, and moreover, the alleged tautomeric shift as a *conditio sine qua non* for H_2 receptor activation was questioned.[20,21] Only the presence of an alkaline nitrogen in the teleposition of the heterocycle appears to be a prerequisite for H_2 stimulation. From the histamine H_2 receptor activity, pK_a values, molecular electrostatic potentials (MEPs) and relative protonation energies for histamine, for a series of (ring) substituted analogues, for thiazoles and for a selenazole compound, it was concluded that only a proton donation from the receptor towards the agonist at the tele position is involved, resulting in a charged ring system and the dicationic species (Fig. 4.1). This is precisely the premise for allowing Timms et al[19] to develop their proton pumping theory for the H_2 receptor.

Also, Spudich[22] speculates that a key function of the seven-transmembrane α-helical motif in both *Archaea* and *Eucarya* is to create the means for proton transfer reactions. This hypothesis is based on a comparison between bacteriorhodopsin (BR), G protein-coupled rhodopsin, and the phototaxis receptor sensory rhodopsin (SR-I) from *Halobacterium salinarium*. Like rhodopsin, SR-I signals to a transducer protein (a G protein in case of rhodopsin), which is complexed with the receptor. In the absence of the transducer system, photoactivation of SR-I results in active proton pumping against a proton gradient; in other words, SR-I then functions as a primary pump. Therefore, the role of the transducer is probably to absorb the proton that is released from the Schiff base upon photoactivation, thereby preventing primary pump activity

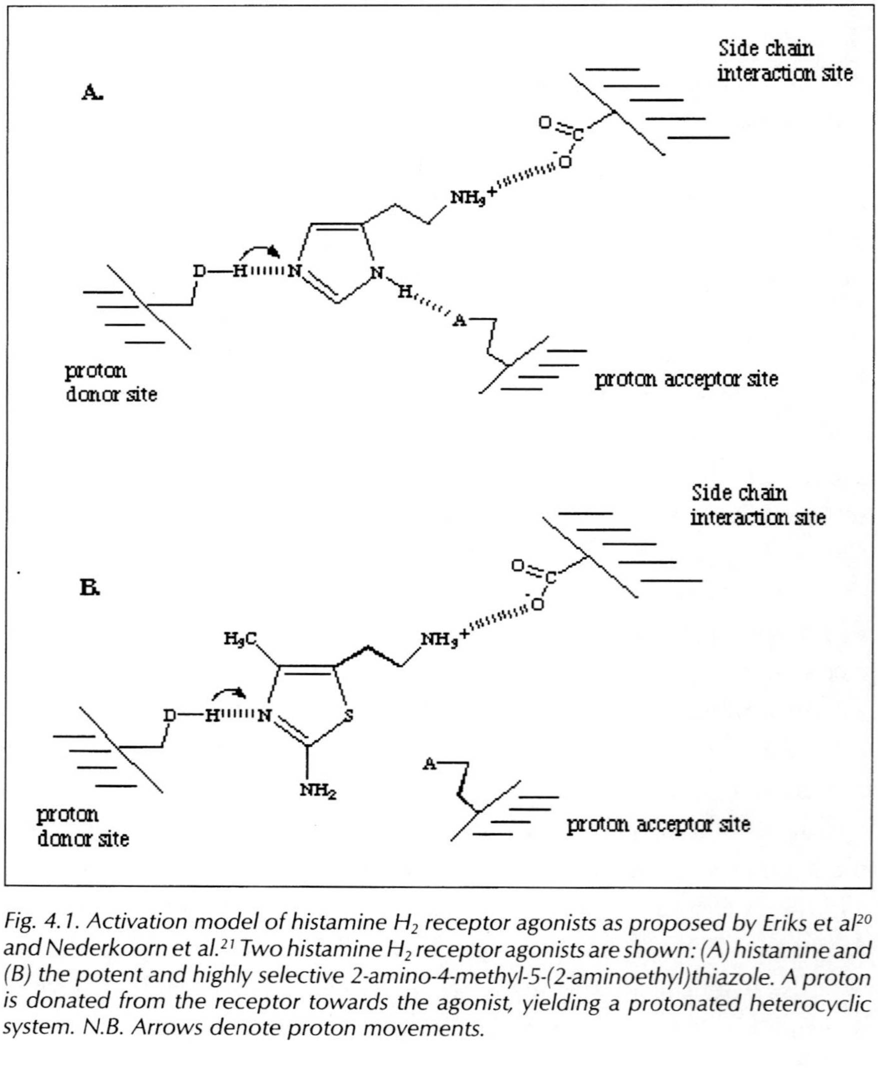

Fig. 4.1. Activation model of histamine H_2 receptor agonists as proposed by Eriks et al[20] and Nederkoorn et al.[21] Two histamine H_2 receptor agonists are shown: (A) histamine and (B) the potent and highly selective 2-amino-4-methyl-5-(2-aminoethyl)thiazole. A proton is donated from the receptor towards the agonist, yielding a protonated heterocyclic system. N.B. Arrows denote proton movements.

of SR-I. The released proton can thus be used in the signal transduction from SR-I to its transducer. Also, rhodopsin activation is associated with proton transfers. In chapter 8, we will address these in more detail.

4.3 TOPIOL'S DELETION MODEL

Topiol[23] postulated a so-called deletion model for the origin of receptors: receptors, together with their agonists (complementary ligands), are considered to be derived from a common parent system. The "parent" system is a collection of entities necessary for the functioning of a complete dynamic biochemical system. Deletion of one entity (α_i subunit) generates an inactive subsystem. Only when the α_i subunit and the inactive subsystem are recombined (to constitute a ligand-receptor pair), does activity return. The relationship between endogenous ligands and biological building blocks, for example histamine and histidine, respectively, or serotonine and tryptophan, respectively, follows directly from this deletion model (Fig. 4.2A). Combining the idea of the ligand-mediated proton pump[18,19] with the deletion model as proposed by Topiol,[23] an amine receptor can be seen as a proton pump that lacks one entity in its proton shuttle in order to pump continuously; agonist binding restores the pumping function (Fig. 4.2B).

From this concept, we inferred a functional analogy with the chemiosmotic theory, which will be presented in detail in chapter 8. We claim to have found Topiol's parent system in the field of bioenergetics.

4.4 CHARACTERISTIC FEATURES OF G PROTEINS

G proteins, also called guanine nucleotide-binding regulatory proteins, play a fundamental role as signal transducers which transmit and modulate signals in cells. They have the ability to receive multiple signals from the exterior, integrate them and activate different cellular amplifier systems and thus control essential life processes in cells (for related reviews in this area see refs. 24-32). The first step in the signalling process involves binding of a ligand to a specific membrane-bound G protein-coupled receptor. The GPCR transmits the signal to the heterotrimeric G protein consisting of three subunits, α, β and γ, inducing the dissociation of the G_α

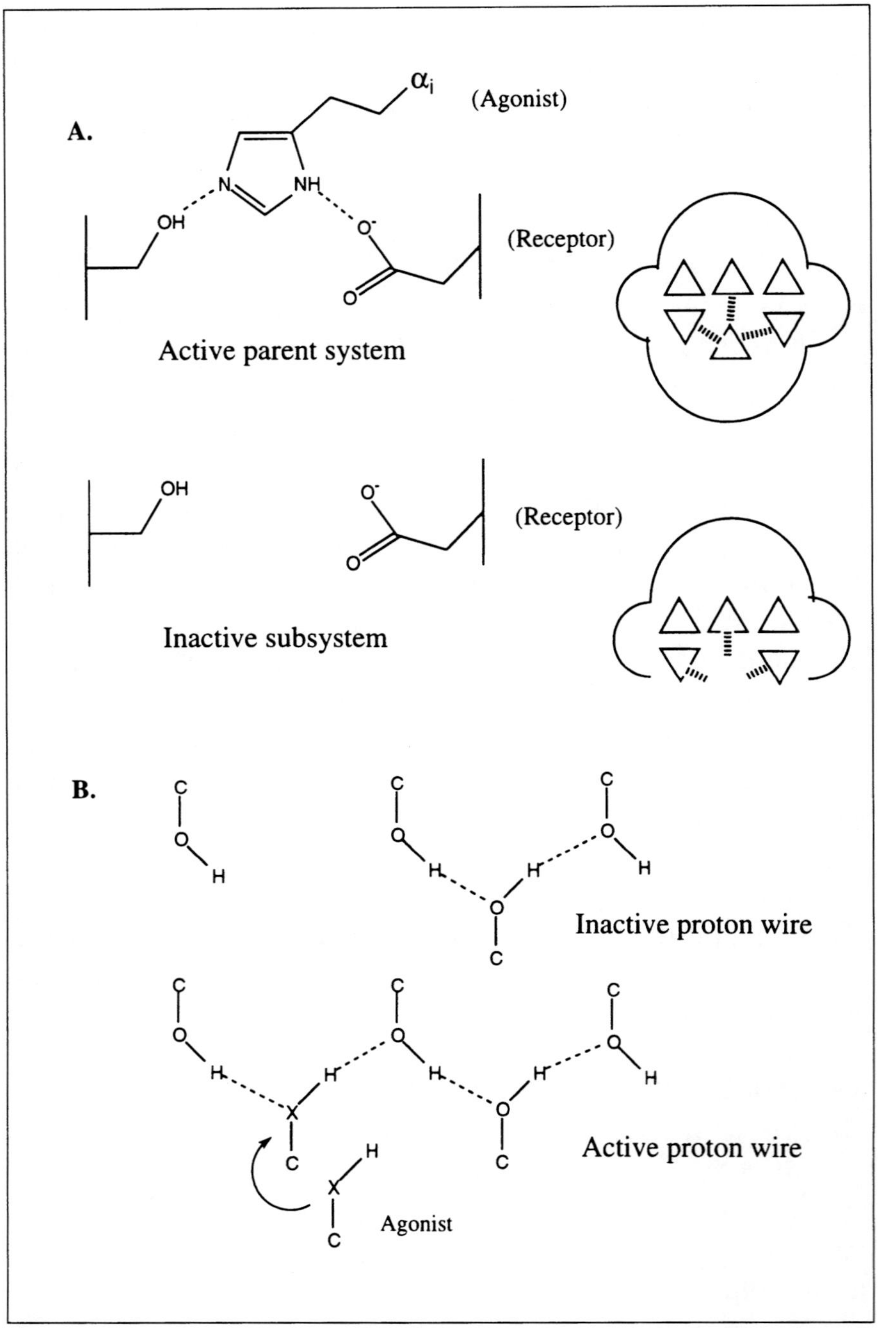

Fig. 4.2. (A) Deletion model for the origin of receptors as suggested by Topiol.[23] As an example of the active parent system, the catalytic triad as found in serine proteases is chosen, illustrating its application to the histamine H_2 receptor (for further details, see ref. 21). (B) Topiol's deletion model with respect to activation possibilities of a hydrogen-bonded chain (HBC). One break in the chain destroys the proton conductance. Ligand binding restores the proton wire and with that the proton translocation abilities.

subunit and thereby transferring the signal further into the cell by influencing an effector system.

In comparison to the number of known GPCRs, the number of G_α subunits involved in receptor-effector coupling is relatively small (about 20).[24,25,29,30] G proteins possess one guanine nucleotide-binding site, although, in the literature, experimental data are reported which should point to the possibility of two nucleotide binding sites.[33-35]

Both the G_α subunit and the $G_{\beta\gamma}$ dimers are effector regulators, either directly or in conjunction with each other.[26,29-31,36,37] $G_{\beta\gamma}$ dimers play a key regulatory role in receptor- and GTPase-dependent shuttling of the G_α subunit between effector and receptor.[25,33,38] The $G_{\beta\gamma}$ dimer seems to increase the affinity of G_α for its receptor. Moreover, it has been suggested that the dimers suppress the 'noise' generated by ligand-unoccupied receptors.[24] At least for rhodopsin, evidence exists that $G_{\beta\gamma}$ binds to the C-terminal region of the GPCR,[39] whereas G_γ appears to determine the G protein's selectivity for its receptor.[40] G_α subunits can be myristoylated or palmitoylated, G_γ can be farnesylated or geranylgeranylated.[41,42] These lipid modifications of G_α, as well as of $G_{\beta\gamma}$, tether heterotrimers to the inner surface of the membrane, thereby enhancing the affinity of G_α for $G_{\beta\gamma}$.

There are at least five homologous G_β subunits and twelve G_γ subunits.[29,30,37,42] Based upon the amino acid similarity of the G_α units, G proteins can be classified into four separate groups (refs. 26,29,43; and refs. *loc. cit.*): (1) G_s (s for stimulatory) mediates, for example, stimulation of adenylate cyclase and the L-type dihydropyridine sensitive Ca^{2+}-channel in skeletal muscle and heart (the Ca^{2+}-channel is closed). Furthermore, cardiac Na^+-channels are reported to be inhibited.[25] The G proteins found in olfactory neuroepithelium (G_{olf}; olf for olfactory) are also classified to belong to the G_s class. All G_s proteins are cholera toxin-sensitive (catalyzed ADP-ribosylation of an Arg residue on G_α, thereby drastically reducing the intrinsic GTPase activity of the G_α subunit; *vide infra*); (2) G_i (i for inhibition) inhibits adenylate cyclase and modulates both K_{ACh} in heart and pituitary cells, the ATP-sensitive K^+-channels in pancreatic β-cells and cardiac ventricular muscle (the K^+-channels are opened). All G_i proteins (i.e., G_{i1}, G_{i2}, G_{i3},

G_g, G_o, G_z, T-r, and T-c) are, except for G_z, pertussin toxin-sensitive (catalyzed ADP-ribosylation of a Cys residue on G_α, thereby blocking receptor-G protein coupling). G_g (g for gusducin) is found in taste buds; at present its precise effector regulation is not known. G_o (o for other) or additional G proteins (sometimes denoted by G_p) stimulate phospholipase C, which releases inositol triphosphate (IP_3) and diacylglycerol (DAG). Also, phospholipase A_2 (hydrolyzing fatty acids at the SN-2 position with subsequent release of arachidonic acid [AA]) is stimulated by a G_o protein. G_z probably inhibits adenylate cyclase, whereas the transducins (T-r from rod cells and T-c from cone cells) stimulate a cGMP-specific phosphodiesterase; (3) The G_q class consisting of G_q, G_{11}, G_{14}, G_{15} and G_{16} stimulates phospholipase C (of the β-type); and (4) the recently discovered G_{12} class consisting of G_{12} and G_{13}, of which the physiological function has not yet been elucidated.[26] Agonists that can activate G_i can override cAMP-dependent actions of agonists that stimulate G_s (for a possible role of $G_{\beta\gamma}$ in this overruling see ref. 38). For a more detailed survey regarding the different subtypes of G proteins, the reader is referred to Birnbaumer and Birnbaumer[26] and to Hepler and Gilman,[43] and for a scheme regarding GPCRs activating different types of G proteins to Birnbaumer.[25]

G proteins have been reported to display several characteristic reactions: (i) dissociation of pre-bound GDP with subsequent binding of GTP followed by dissociation of the G_α subunit from the G protein and yielding an activated $G_\alpha{}^*{\cdot}$GTP complex plus the dimeric protein $G_{\beta\gamma}$; and (ii) in situ hydrolysis of GTP (GTPase activity) which eventually leads to reassociation of $G_{\beta\gamma}$ with G_α. In the test-tube this reaction cycle proceeds slowly if at all. However, after adding Mg^{2+} and non-hydrolyzable GTP analogues (guanosine 5'-[β,γ-imido] triphosphate (p[NH]ppG) or guanosine 5'-[γ-thio]phosphate [GTPγS]), the G_α dissociation step in the cycle proceeds readily.[25,28]

In addition to the aforementioned two characteristic reactions, other phenomena in relation to G protein activation have recently been reported, i.e., Mg^{2+}-dependent nucleoside diphosphate kinase activity of the trimeric G protein.[26,35,44-47] Nucleoside diphosphate kinases (NDPKs) catalyze the phosphorylation of nucleoside diphosphates (NDPs) to nucleoside triphosphates (NTPs) via a transiently phosphorylated enzyme intermediate in a Mg^{2+}-dependent way. Evidence is available that NDPKs are involved in the activation of G proteins, and it has been suggested that in this way NDPKs provide a local supply of GTP.[48] Now, it has actually been concluded that the trimeric G protein itself becomes phosphorylated during (Mg^{2+}-dependent) NTP hydrolysis.[26,35,44-47] A histidine residue on the G_β subunit is phosphorylated and presumably the trimeric G protein itself acts as a NDPK catalyzing GTP synthesis from GDP and P_i. Moreover, agonist-activated receptors are reported to stimulate the overall GTP synthesis (NDPK) activity.[35] Also GTPγS is found to contribute to transphosphorylation reactions. The histidine residue on the G_β subunit can be thiophosphorylated and subsequently GDP is found to be converted into GTPγS, leading to a constantly activated G_α*·GTPγS. In contrast, p[NH]ppG does not seem to be involved in transphosphorylation; it is known that, at equipotent concentrations, receptor-stimulated maximal GTPγS binding to the G protein is about 2-fold higher than that of p[NH]ppG.[35]

4.5 STRUCTURAL FEATURES OF G PROTEINS

Before any X-ray information became available, Conklin and Bourne[11] derived a structural model for G_α based on data from biochemical, immunological, and molecular genetics studies. Also the high primary sequence identity between diverse G_α units (50-90%) and members of the GTPase superfamily ($p21^{ras}$ and elongation factor Tu) was considered. The model encompasses surfaces for binding to GPCRs (C-terminal region of G_α), effectors (most important T-r_α residues are 293-314), $G_{\beta\gamma}$ dimers (N-terminal region of G_α), and fatty acids (N-terminal region). Fatty acids (*vide supra*) are known to improve the binding of G_α to both the membrane and $G_{\beta\gamma}$.[11,26,41] Three different conformations for G_α monomers are predicted: inactive G_α with bound GDP, active G_α* with bound GTP and a transition state with an empty nucleotide binding site occurring after GDP release and before GTP coupling.

Shortly after this model appeared in the literature, an X-ray structure of T-r_α* complexed with GTPγS became available.[49] Recently, also the three-dimensional structure of inactive T-r_α

complexed with GDP was also reported.[50] G_α appears to have two domains, i.e., a GTPase and an α-helical domain with the guanine nucleotide bound in a cleft in between these two domains (cf. Fig. 4.3A and B). A comparison of the $T\text{-}r_\alpha{}^*$ and $T\text{-}r_\alpha$ structures reveals that the conformational changes occurring upon activation are restricted to three regions, which are therefore called switch regions.[50] Structural changes in switch region I (Ser^{173}-Thr^{183}; if not stated otherwise G protein amino acid numbering refers to $T\text{-}r_\alpha$ throughout this chapter) are due to H-bond formation between the γ-phosphate of GTP, Thr^{177} and the cholera toxin-sensitive Arg^{174}. Switch region II (Phe^{195}-Thr^{215}) is both stretched and rotated in the GTPγS bound form. This rotation, occurring upon G protein activation, is caused by a rupture of the H-bond between His^{209} and Glu^{212}, which enhances the rotational freedom of the α_2 helix in G_α. In the GDP bound form, both Arg^{201} and Arg^{204} are solvent-exposed and rotate to an interior position in the activated structure. Both arginines can then bind to Glu^{241} of the α_3 helix via a salt bridge. The changes in switch regions I and II are mediated through H-bonds. Also changes in switch region III (Asp^{227}-Arg^{238}) are suggested to result from altered hydrogen-bonded networks between these three regions. Formation of H-bonds between the carboxylate group of Glu^{232} and the backbone amide of Arg^{201} and the guanidinium of Arg^{204} (water-mediated) are thought to be of great importance for the activation process. Switch region II is (next to the above mentioned N-terminus, and the region denoted by i_1)[49,50] proposed to play a role in binding the $G_{\beta\gamma}$ dimer (refs. 30,50; and refs. *loc. cit.*). Switch regions I and II are not only found in G proteins, but also in the Ras superfamily (e.g., $p21^{ras}$) and elongation factors (e.g., EF-Tu), whereas switch region III seems to be unique to G proteins.

Noel et al[49] used the above structural information to explain how $G_\alpha{}^*$ is able to hydrolyze GTP but not GTPγS. They proposed a mechanism very similar to the ATP hydrolysis mechanism as described for ATP synthases in chapter 3. Within the active site of $G_\alpha{}^*$, a water molecule is positioned close to the γ-phosphorus group. At first, a glutamic acid (Glu^{203} in $T\text{-}r_\alpha{}^*$) was assumed to activate the water molecule, resulting in a hydrolytic attack on GTP. An arginine (Arg^{174} in $T\text{-}r_\alpha{}^*$) was thought to prevent the

γ-thiophosphate group from adopting the transition state needed for hydrolysis. However, Coleman et al[51] elucidated the three-dimensional structure of $G_{i\alpha}$* complexed with GDP·AlF^{4-}, a transition state analogue of GTP. In $G_{i\alpha}$, both Arg^{178} (corresponding to Arg^{174} in T-r) and Gln^{204} (analogue to Gln^{200} in T-r_{α}) seem to stabilize the transition state structure. The $G_{i\alpha}$'s Gln^{204} is believed to polarize the putative nucleophilic water molecule rather than Glu^{207} (corresponding to Glu^{203} in T-r). Kleuss et al,[52] however, rejected the possibility of Glu^{207} playing a key role in $G_{i\alpha}$'s hydrolysis activity. Although Glu^{207} is highly conserved in G_{α} subunits, mutation to either an Ala or Gln did not reveal an important role for this residue in hydrolysis activity; in contrast, mutation of Gln^{204} impairs GTPase activity (ref. 52, and refs. *loc. cit.*). The important role of a glutamine residue in T-r_{α} (Gln^{200}, instead of the earlier proposed Glu^{203}) for stabilizing the transition state of the hydrolysis reaction has now also been established.[53]

For a detailed review concerning structural and functional relationships between G proteins, the reader is referred to Rens-Domiano and Hamm,[37] who indicate that the key difference between the G_{α} and G_{α}* state is the extent of the exposure of the nucleotide to the solvent. In both conformations, the ribose and guanine ring are shielded from the solvent. Within G_{α}, the nucleotide binding site is more open with the β-phosphate and Mg^{2+} surface-exposed, whereas the nucleotide cleft in G_{α}* is more closed and the γ-phosphate and Mg^{2+} are shielded from the solvent.

Very recently, crystal structures of trimeric $G_{\alpha\beta\gamma}$ as well as dimeric $G_{\beta\gamma}$ were published (cf. Fig. 4.3 A-C).[54-56] The $G_{\alpha i1\beta 1\gamma 2}$ was resolved at 2.3 Å[56] and the chimeric $G_{\alpha t/i1\beta t\gamma t}$ at 2.0 Å.[54] The G_{α} subunit of the latter consists of T-r_{α} with its residues 216-294 replaced with the corresponding residues (220-298) from $G_{\alpha i1}$. The dimer T-$r_{\beta\gamma}$ was resolved at 2.1 Å.[55] The crystallized complexes are intact but lipid or other covalent modifications as well as water molecules are absent. The heterotrimer is asymmetric. An extended interface between G_{β} and nearly all of G_{γ} is observed. The G_{β} subunit makes localized contacts with G_{α}, which involve the aforementioned switch II (so-called "switch interface") and the helical amino terminal region of G_{α} (so-called "N-terminal interface"). Lambright et al[55] report additional interactions between residues

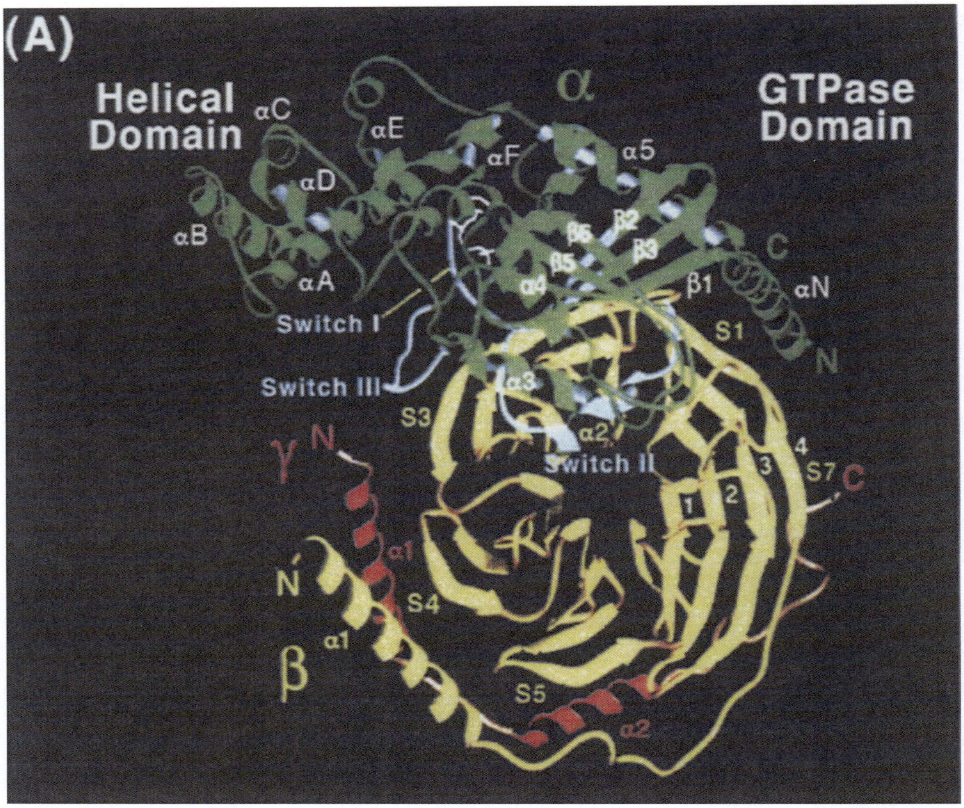

Fig. 4.3. Ribbon drawings of the structure of a heterotrimeric G protein. (A) The G_α subunit is shown in green, G_β in yellow and G_γ in red. The GDP nucleotide is colored white and the switch regions I-III cyan. View is down the axis of the G_β propeller domain. (B) View rotated 70° about the horizontal axis compared with A. (C) Polypeptide chains connecting the WD40 cores are colored blue, containing the D strands of each β blade. The strands of blade 1 are labelled (A-D) and each blade is numbered (1-7). The view is taken from G_β looking towards G_α. The backbone atoms of the switch II of G_α and selected side chains are colored red. (Figures 4.3A and B are reprinted with permission from: Lambright DG, Sondek J, Bohm A et al. The 2.0 Å crystal structure of a heterotrimeric G protein. Nature 1996; 379:311-319.) (Figure 4.3C is reprinted with permission from: Wall MA, Coleman DE, Lee E et al. The structure of the G protein heterotrimer $G_{i\alpha1\beta1\gamma2}$. Cell 1995; 83:1047-1058.) (Please see color insert.)

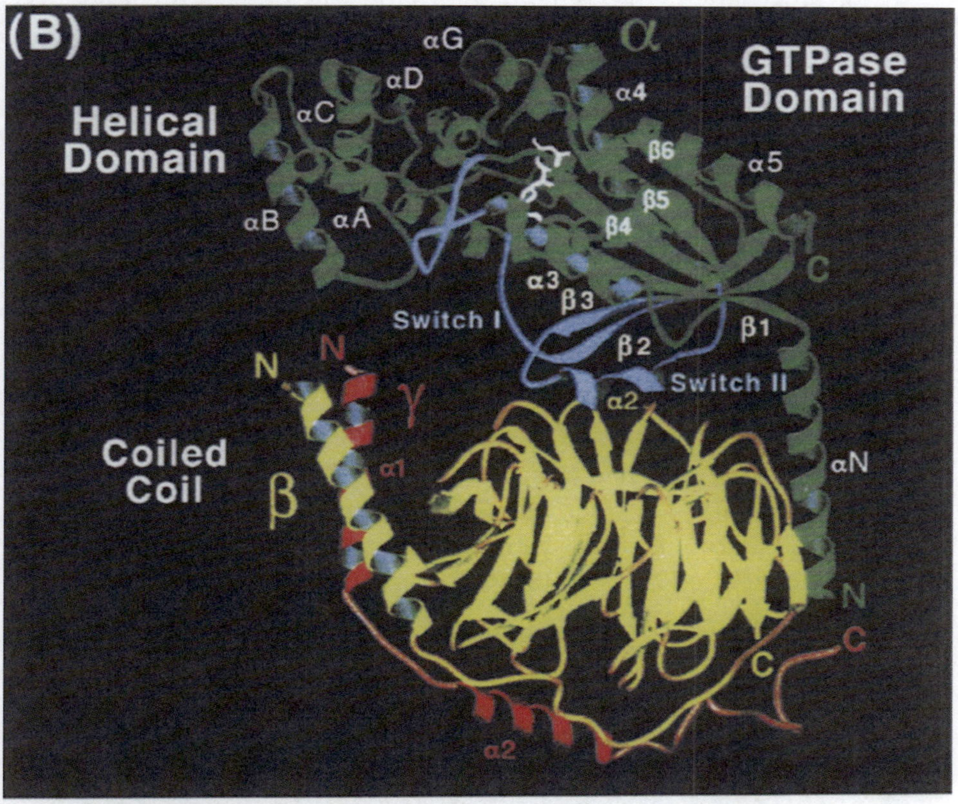
(B)
αG
α
GTPase
Domain
αD
α4
Helical
Domain
αC
β6
α5
β5
αB
αA
β4
C
α3
β3
Switch I
β1
β2
N
N
Switch II
γ
α2
Coiled
Coil
α1
αN
β
N
C
C
α2

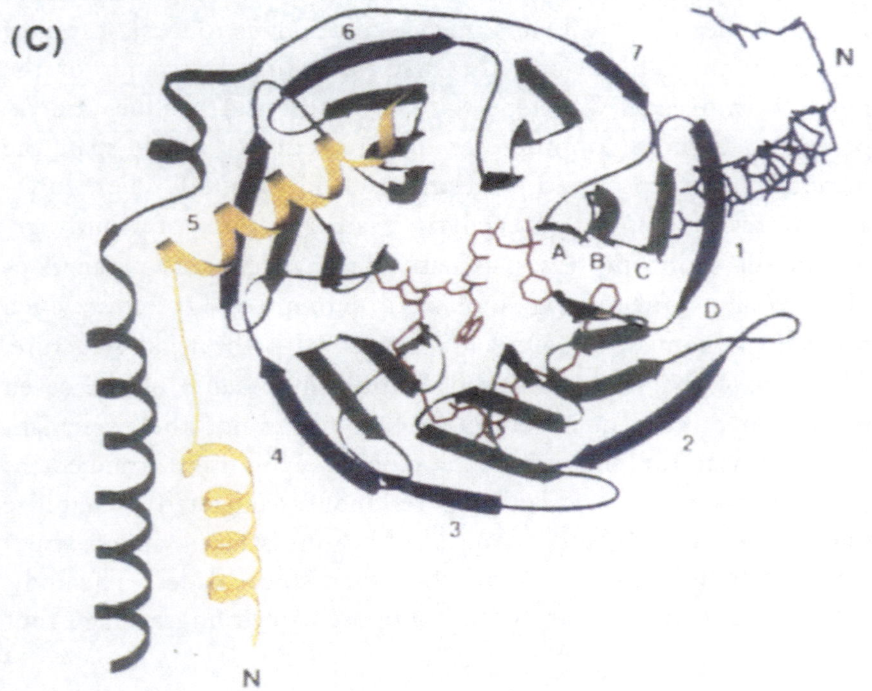
(C)
6
7
N
5
A
B
C
1
D
2
4
3
N
N

in, or adjacent to, the switch I and II regions of G_α and residues from loops and turns at the top of the β-propeller domain of G_β. The extended helical N-terminus of G_α, crucial to trimer formation, is stabilized by the interactions at the N-terminal interface. Deletion of the G_α N-terminus prevents trimer formation.[56]

G_β subunits belong to the so-called WD(40) family (W for tryptophan; D for aspartic acid). This family is characterized by WD-repeat motifs of approximately 40 amino acids long and containing a number of conserved amino acids including a Trp-Asp dipeptide (i.e., WD frequently terminate the repeat and sets of 4-8 repeats are normally found). Such WD repeats have been determined in the sequences of about 40 eukaryotic proteins other than G_β (see ref. 55 and refs. *loc. cit.*). The G_β is the first protein from this family to have its three-dimensional structure experimentally determined. The G_β subunit displays a toroidal β-propeller domain composed of seven motifs (for examples of other crystallized β-propeller-containing proteins the reader is referred to refs. 55,56 and to refs. therein). Each of the seven motifs (β-propeller) comprises a four-stranded antiparallel β-sheet, which is designated a "β-blade."[56] The approximate 7-fold symmetry is reflected in the amino acid sequence, consisting of seven WD repeats. It is noted that a single WD repeat does not correspond to a single β-sheet. The WD40 sequence corresponds to the last (outer) strand of one β-blade and the first three (inner) strands of the next. Within each WD repeat in G_β, conserved residues are reported to form a common set of interactions (which span the breadth of the repeat and position adjacent β-blades). These interactions reveal a so-called catalytic triad structure, previously reported for serine proteases. A catalytic triad consists of an Asp-His-Ser/Thr motif (cf. Fig. 4.2A). Furthermore, a Trp (after which the WD is named) is found to interact with the triad structure. The Asp of the structural triad is the only residue of the seven repeats that is fully conserved (N.B. This is not the trademark Asp found at the end of a WD repeat). This triad arrangement (and especially the invariant Asp residue) is suggested to stabilize the β-blade structure and is conserved in four blades. Modest structural differences are found for the other three blades. The individual WD40 motifs can be superimposed, revealing average root mean square deviations between corresponding C_α atoms of 0.6-1.2 Å.[56] The seven β-blades form a narrow central channel with an average diameter (from C_α to C_α) of approximately 12 Å. G_α is positioned at the narrow end of this channel (Fig. 4.3A-C).

The interactions between G_α and G_β at the switch and N-terminal interface G_α are mainly of hydrophobic nature. Although the switch II residues 203-209 of $G_{\alpha i1}$ are located within the switch interface, the side chains of these residues do not form direct contacts with G_β. Which precise combinations of the different types of the three subunits α, β and γ are possible have yet to be clarified.[56] In the heterotrimer, the carboxyl terminus of G_α is exposed for possible interactions with the appropriate GPCRs. Direct interaction of a receptor with switch II is proposed by Wall et al.[54] This is in accordance with a theoretical model for a ligand-receptor-G_α protein complex, which we had proposed before any structural information on $G_{\alpha\beta\gamma}$ or $G_{\beta\gamma}$ became available.[57] Access of the appropriate intracellular part of a receptor to switch II of G_α could actually occur via the cleft between the G_α and G_β subunits at the switch interface. The negative electrostatic surface in this region (mainly caused by G_β) is suggested to provide sites of interaction with cationic receptor groups.[56] The apposition of the two subunits (i.e., G_α and G_β, Fig 4.3C, red) in this region shields this vital domain, protecting the G_α face until the cationic lever of the receptor 'flips' the switch domain and triggers the G_α conformational change.[42] (N.B. For a possible molecular mechanism see ref. 57).

The N-terminus of G_α is known to interact with receptors (ref. 56, and refs. therein). This means that these receptor-G_α interactions displace the G_α-G_β interactions at the N-terminal interface and thus promote subunit dissociation. As indicated, the transition of G_α to G_α* includes the (120°) rotation of the α_2 helix from the switch II region. This rotation destabilizes the interactions at the switch interface between G_α and G_β. Furthermore, the same rotation creates two ionic interactions between G_α* and G_β: $G_{\alpha i1}$ Glu^{216} forms a salt bridge with $G_{\beta 1}$ Lys^{57} and $G_{\alpha i1}$ Lys^{210} is inserted into a negatively charged pocket formed by Asp^{228} and Asp^{246} on adjacent loops of $G_{\beta 1}$.[56] It is noted that the $G_{\beta\gamma}$ functions as a rigid unit since its structure does not significantly differ in the trimer or dimer form.[55] The presence of two nucleotide binding sites as earlier proposed[33-35] has not been validated by the aforementioned X-ray structures.

4.6 SUMMARY

Hundreds of membrane-bound receptors use G proteins to transduce an external signal into the cell. The membrane-spanning portion of these G protein-coupled receptors consists of seven (almost antiparallel) α-helices. The binding of a ligand (external signal) occurs in between the transmembrane (and sometimes extracellular) domains of the GPCR. Inside the cell, G proteins are membrane-attached and the coupling of the G protein to the appropriate GPCR is mediated via the receptor's intracellular loops and C-terminus. Upon ligand binding, the GPCR is activated, thereby passing on a signal to the G protein, which in turn becomes activated itself. For an extended survey concerning the molecular biology of GPCRs, the reader is referred to Iismaa et al.[58]

The activation of the G protein involves two characteristic reactions: (i) dissociation of pre-bound GDP with subsequent binding of GTP followed by the dissociation of the G_α subunit from the heterotrimeric $G_{\alpha\beta\gamma}$ protein, yielding an activated $G_\alpha{}^*$·GTP complex and the dimeric $G_{\beta\gamma}$ protein; and (ii) in situ hydrolysis of GTP at the $G_\alpha{}^*$·GTP complex which eventually leads to reassociation of $G_{\beta\gamma}$ with G_α. Recently, a third characteristic reaction concerning G protein activation has been reported: (iii) heterotrimeric G proteins are found to display nucleoside diphosphate kinase activity, i.e., G proteins are able to catalyze the phosphorylation of GDP to GTP. Ligand-activated GPCRs stimulate this GTP synthesis.

Based upon amino acid similarity of G_α subunits, G proteins are classified into four separate groups (G_s, G_i, G_q and G_{12}). G proteins within each class regulate different second messenger systems. Three-dimensional X-ray structures of $G_{\alpha\beta\gamma}$, $G_{\beta\gamma}$, G_α·GDP and $G_\alpha{}^*$·GTP reveal conformational changes occurring G protein activation. So far, experimentally determined three-dimensional structures for GPCRs have not been reported. From two-dimensional images, it is clear that GPCRs have a common heptahelical (transmembrane) motif. How exactly the receptor couples to a

G protein, how ligands bind and subsequently can induce G protein activation and how G_α*·GTP couples to effector molecules still remain questions to be answered. Eventually new X-ray structures of these complexes and modelling studies will aid in elucidating the underlying molecular mechanisms. In the meantime, theoretical studies have revealed that proton transfers could be an integral part of the receptor's signal transduction process. A deletion theory for the origin of receptors has been proposed in which receptors are considered as a collection of entities from which one crucial part is missing for function as an active biological system. The ligand is regarded as the missing part, so that binding of the ligand restores the biological function.

References

1. Barnard EA. Receptor classes and the transmitter-gated ion channels. Trends Biochem Sci 1992; 17:368-374.
2. Findlay JBC, Pappin DJC. The opsin family of proteins. Biochem J 1986; 238:625-642.
3. Hibert MF, Trumpp-Kallmeyer S, Hoflack J et al. This is not a G protein-coupled receptor. Trends Pharmacol Sci 1993; 14:7-12.
4. Henderson R, Baldwin JM, Ceska TA et al. Model for the structure of bacteriorhodopsin based on high-resolution electron cryo-microscopy. J Mol Biol 1990; 213:899-929.
5. Schertler GFX, Villa C, Henderson R. Projection structure of rhodopsin. Nature 1993; 362:770-772.
6. Baldwin JM. The probable arrangement of the helices in G protein-coupled receptors. EMBO J 1993; 12:1693-1703.
7. Unger VM, Schertler GFX. Low resolution structure of bovine rhodopsin determined by electron cryo-microscopy. Biophys J 1995; 68:1776-1786.
8. Hoflack J, Trumpp-Kallmeyer S, Hibert M. Re-evaluation of bacteriorhodopsin as a model for G protein-coupled receptors. Trends Pharmacol Sci 1994; 15:7-9.
9. Donnelly D, Findlay JBC, Blundell TL. The evolution and structure of aminergic G protein-coupled receptors. Receptors and Channels 1994; 2:61-78.
10. Collins S. Molecular structure of G protein-coupled receptors and regulation of their expression. Drug News & Perspectives 1993; 6:480-487.
11. Conklin BR, Bourne HR. Structural elements of G_α subunits that interact with $G_{\beta\gamma}$, receptors and effectors. Cell 1993; 73:631-641.

12. Lee NH, Kerlavage AR. Molecular biology of G protein-coupled receptors. Drug News & Perspectives 1993; 6:488-497.
13. Oliveira L, Paiva ACM, Vriend G. A common motif in G-protein-coupled seven transmembrane helix receptors. J Comp-Aided Mol Design 1993; 7:649-658.
14. Ganellin CR. Chemical constitution and prototropic equilibria in structure-activity analysis. In: Roberts GCK, ed. Drug action at the molecular level. London: MacMillan Press Limited, 1977:1-39.
15. Weinstein H, Chou D, Johnson CL et al. Tautomerism and the receptor action of histamine: A mechanistic model. Mol Pharmacol 1976; 12:738-745.
16. Weinstein H, Mazurek AP, Osman R et al. Theoretical studies on the activation mechanism of histamine H_2 receptor: The proton transfer between histamine and a receptor model. Mol Pharmacol 1986; 29:28-33.
17. Osman R, Topiol S, Rubinstein L et al. A molecular model for activation of a 5-hydroxytryptamine receptor. Mol Pharmacol 1987; 32:699-705.
18. Timms D, Wilkinson AJ, Kelly DR et al. Interactions of Tyr^{377} in a ligand-activation model of signal transmission through β_1 adrenoceptor α-helices. Int J Quant Chem: Quant Biol Symp 1992; 19:197-215.
19. Timms D, Wilkinson AJ, Kelly DR et al. Ligand-activated transmembrane proton transfer in β_1 adrenergic and m_2 muscarinic receptors. Receptors and Channels 1994; 2:107-119.
20. Eriks JCh, v.d. Goot H, Timmerman H. New activation model for the histamine H_2 receptor, explaining the activity of the different classes of histamine H_2 receptor agonists. Mol Pharmacol 1993; 44:886-894.
21. Nederkoorn PHJ, Vernooijs P, Donné-Op den Kelder GM et al. A new model for the agonistic binding site on the histamine H_2 receptor: The catalytic triad in serine proteases as a model for the binding site of histamine H_2 receptor agonists. J Mol Graphics 1994; 12:242-256.
22. Spudich JL. Protein-protein interaction converts a proton pump into a sensory receptor. Cell 1994; 79:747-750.
23. Topiol S. The deletion model for the origin of receptors. Trends Biochem Sci 1987; 12:419-421.
24. Birnbaumer L, Mattera R, Yatani A et al. Recent advances in the understanding of multiple roles of G proteins in coupling of receptors to ionic channels and other effectors. In: Moss J, Vaughan M, eds. ADP-ribosylating toxins and G proteins, Insights into signal transduction. Washington DC: Am Soc Microbiol 1990:225-266.

25. Birnbaumer L. G proteins and the modulation of potassium channels. In: Weston AH, Hamilton ThC, eds. Potassium channel modulators. Oxford: Blackwell Scientific Publications, 1993:44-75.
26. Birnbaumer L, Birnbaumer M. Signal transduction by G proteins: 1994 edition. J Receptor & Signal Transduction Research 1995; 15:213-252.
27. Chabre M, Antonny B, Vuong TM. The transducin cycle in the phototransduction cascade. NATO ASI Series, Series H 1991; 52:207-220.
28. Gilman AG. G proteins: Transducers of receptor-generated signals. Annu Rev Biochem 1987; 56:615-649.
29. Gilman AG. G-Proteine und die Regulation der Adenylat-Cyclase (Nobel-Vortrag). Angew Chem 1995; 107:1533-1548.
30. Müller S, Lohse MJ. The role of G-protein βγ subunits in signal transduction. Biochem Soc Trans 1995; 23:141-148.
31. Neer EJ. Heterotrimeric G proteins: organizers of transmembrane signals. Cell 1995; 80:249-257.
32. Stryer L. Cyclic GMP cascade of vision. Ann Rev Neurosci 1986; 9:87-119.
33. Godchaux W III, Zimmerman WF. Membrane-dependent guanine nucleotide binding and GTPase activities of soluble protein from bovine rod cell outer segments. J Biol Chem 1979; 254:7874-7884.
34. Fung BK-K, Stryer L. Photolyzed rhodopsin catalyzes the exchange of GTP for bound GDP in retinal rod outer segments. Proc Natl Acad Sci USA 1980; 77:2500-2504.
35. Kaldenberg-Stasch S, Baden M, Fesseler B et al. Receptor-stimulated guanine nucleotide-triphosphate binding to guanine nucleotide-binding regulatory proteins. Eur J Biochem 1994; 221:25-33.
36. Clapham EH, Neer EJ. New roles for G-protein βγ-dimers in transmembrane signalling. Nature 1993; 365:403-406.
37. Rens-Domiano S, Hamm HE. Structural relationships of heterotrimeric G-proteins. FASEB J 1995; 9:1059-1066.
38. Manning DR. Regulatory role of the βγ subunits. In: Moss J, Vaughan M, eds. ADP-ribosylating toxins and G proteins, Insights into signal transduction. Washington DC: Am Soc Microbiol, 1990:349-370.
39. Philips WJ, Cerione RA. Rhodopsin/transducin interactions. I. Characterization of the binding of the transducin-βγ subunit complex to rhodopsin using fluorescence spectroscopy. J Biol Chem 1992; 267:17032-17039.
40. Kisselev O, Gautam N. Specific interaction with rhodopsin is dependent on the γ subtype in a G protein. J Biol Chem 1993; 268:24519-24522.
41. Casey PJ. Protein lipidation in cell signalling. Science 1995; 268:221-225.

42. Clapham DE. The G protein nanomachine. Nature 1996; 379:297-299.
43. Hepler JR, Gilman A. G proteins. Trends Biochem Sci 1992; 17:383-387.
44. Klinker JF. Transducin possesses NTPase activity. Naunyn-Schmiedeberg's Arch Pharmacol 1995; 351:R68.
45. Seifert R. Transducin is a nucleoside diphosphate kinase. Naunyn-Schmiedeberg's Arch Pharmacol 1995; 351:R68.
46. Wieland T, Kaldenberg-Stasch S, Fesseler B et al. Regulation of G protein function by phosphorylation. Can J Physiol Pharmacol 1994; 72:S5.
47. Wieland T, Nürnberg B, Ulibarri I et al. Guanine nucleotide-specific phosphate transfer by guanine nucleotide-binding regulatory protein β-subunits. J Biol Chem 1993; 268:18111-18118.
48. Klinker JF, Hagelüken A, Grünbaum L et al. Mastoparan may activate GTP hydrolysis by G_i proteins in HL-60 membranes indirectly through interaction with nucleoside diphosphate kinase. Biochem J 1994; 304:377-383.
49. Noel JP, Hamm HE, Sigler PB. The 2.2 Å crystal structure of transducin-α complexed with GTPγS. Nature 1993; 366:654-663.
50. Lambright DG, Noel JP, Hamm HE et al. Structural determinants for activation of the α-subunit of a heterotrimeric G protein. Nature 1994; 369:621-628.
51. Coleman DE, Berghuis AM, Lee E et al. Structures of active conformations of $G_{i\alpha 1}$ and the mechanism of GTP hydrolysis. Science 1994; 265:1405-1412.
52. Kleuss C, Raw AS, Lee E et al. Mechanism of GTP hydrolysis by G protein α subunits. Proc Natl Acad Sci USA 1994; 91:9829-9831.
53. Sondek J, Lambright DG, Noel JP et al. GTPase mechanism of G proteins from the 1.7Å crystal structure of transducin $\alpha \cdot GDP \cdot AlF_4^-$. Nature 1994; 372:276-279.
54. Wall MA, Coleman DE, Lee E et al. The structure of the G protein heterotrimer $G_{i\alpha 1\beta 1\gamma 2}$. Cell 1995; 83:1047-1058.
55. Lambright DG, Sondek J, Bohm A et al. The 2.0Å crystal structure of a heterotrimeric G protein. Nature 1996; 379:311-319.
56. Sondek J, Bohm A, Lambright DG et al. Crystal structure of a G_A protein βγ dimer at 2.1 Å resolution. Nature 1996; 379:369-374.
57. Nederkoorn PHJ, Timmerman H, Donné-Op den Kelder GM et al. GTP synthases. Proton pumping and phosphorylation in G_α protein-receptor complexes. Receptors & Channels, in press.
58. Iismaa TP, Biden TJ, Shine J. In: G protein-coupled receptors. Austin: RG Landes Co., 1995.

CHAPTER 5

The (Extended) Ternary Complex Model ([E]TCM) for G Protein Activation

5.1 THE CLASSICAL THEORETICAL FRAMEWORK

Many reviews have been written in which the mechanism of G protein activation is described. Several authors including Birnbaumer,[1-3] Chabre et al,[4] Conklin and Bourne,[5] Gilman,[6,7] Rodbell[8] and Stryer[9] describe in detail how G proteins may be activated by GPCRs; extensive lists of references are presented *loc. cit.* In this subsection we summarize the most important findings.

The classical mechanism for G protein activation, the so-called ternary complex model (TCM), involves only three species: ligand (H[ormone]), receptor (R) and G protein (G).[10,11] Upon activation of a GPCR by a ligand (chemical substance or light), the GDP-bound form of the G protein binds to the receptor resulting in an exchange of GDP for GTP. This is followed by dissociation of the trimeric G protein, giving an activated $G_\alpha{}^*$·GTP subunit and the $G_{\beta\gamma}$ dimer. $G_\alpha{}^*$·GTP is able to activate an effector system and after subsequent GTP hydrolysis the G_α subunit reassociates with the $G_{\beta\gamma}$ dimer, yielding the original trimer (cf. Fig. 5.1). However, Samama et al[12] report findings which cannot be explained within this "classical" model.

Samama et al[12] (and refs. *loc. cit.*) report a mutation on the β_2 adrenergic receptor that leads to agonist-independent activation of adenylate cyclase (the effector), in other words "basal activity" is induced or at least increased. In addition, this so-called constitutively active mutant receptor shows: (i) an increased potency of agonists for stimulation of the effector system; (ii) an increased affinity for agonists, even in the absence of G protein, but not for antagonists. The extent of increase of affinity correlates with the intrinsic activity of the ligand; and (iii) an increased intrinsic activity of partial agonists. These findings—particularly the efficacy-related change in affinity for the uncoupled state of the mutant receptor—cannot be explained within the classical theoretical framework and therefore the model was extended.[12] The next subsections summarize the main features of this extended model (ETCM).

5.2 COMPONENTS OF THE ETCM

Within the ETCM the different molecular species are: H, R, R*, G, HR, HR*, R*G and HR*G. Note that RG and HRG are not present. So, the authors[12] introduce the presence of two receptor states, R and R*. For a review regarding the two-state model of receptor activation and the underlying mathematics, the reader is referred to Leff.[13] It is assumed that only the activated receptor R* is able to couple to the G protein, so that HR*G is the only possible ternary complex to be formed (Fig. 5.1). At physiological GTP concentrations, both the ternary complex and the trimeric $G_{\alpha\beta\gamma}$ form are ephemeral since GTP is thought to bind instantaneously to the G_α subunit, leading to dissociation of G_α*·GTP from the ternary complex and subsequently to the transfer of R* to R. G_α*·GTP associates with an effector protein for which it has high affinity. Effector stimulation is terminated by intrinsic GTPase activity of G_α*·GTP. Subsequently, the G_α subunit reassociates with the $G_{\beta\gamma}$ dimer, once again able to form another HR*G ternary complex. Neither the classical nor the extended model include an explicit role for GTP.[12] The theoretical consequences of the ETCM are discussed in the following subsections.

It is noteworthy that Shankley and co-workers interpret the ETCM in a sense that the R* state of a GPCR can be seen as the situation in which a GPCR is coupled to a G protein, whereas the R state is considered as the uncoupled state. Consequently, in this latter model the HR* state does not exist, as we have seen for the non-existent states RG and HRG as indicated by Samama et al.[12] Shankley (Nigel P. Shankley, personal communication) claims that under physiological circumstances the mathematics underlying the

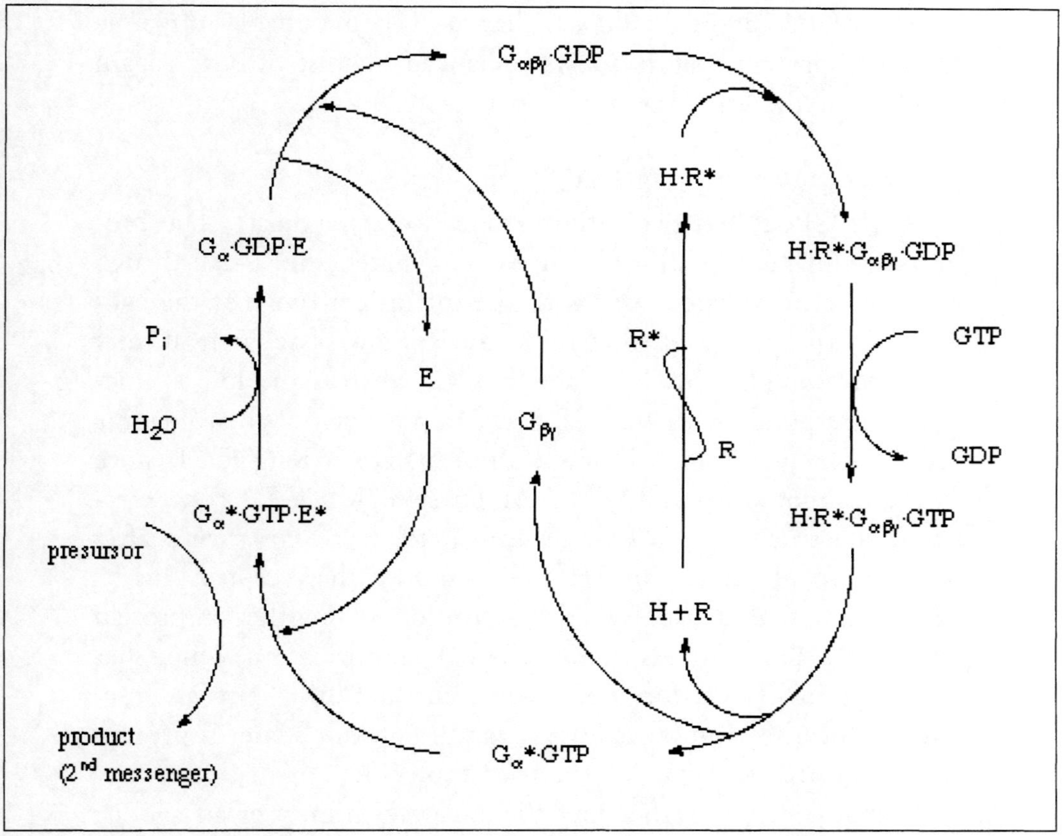

Fig. 5.1. Activation mechanism for G proteins (adapted from Gilman[6]) applied to the extended ternary complex model (ETCM): only activated receptors (R) are seen to bind/activate G proteins.*

ETCM[12,13] and his model, in which HR* does not appear, can both account for the pharmacological behavior of agonists as well as of antagonists. This implies that, if the HR* state would exist at all (e.g., refs. 14,15; *vide infra*), its total amount under normal physiological conditions must be very moderate with respect to the other states in the ETCM. Normally it is found that inclusion of approximately 100 μM of the non-hydrolyzable GTP analogue p[NH]ppG (thereby preventing G protein coupling to a GPCR) converts all agonist binding into the low affinity form (i.e., HR; *vide infra*), which is in accordance with the conception of R* as the coupled state. However, Leeb-Lundberg et al, who studied the B_2 bradykinin receptor, still find a considerable level of high affinity binding in the presence of pp[NH]pG, indicating that the HR* state does exist.[14] Furthermore, Strader and co-workers[15] report a mutant of a β-adrenergic GPCR that has become unable to couple to the G protein, yet it shows enhanced agonist affinity, again pointing at the existence of the HR* state.

Agonists Within the ETCM

Each GPCR has two affinity states for its agonists. The proportion and relative affinity of these two states, the so-called high and low affinity states, vary with the intrinsic activity of the agonist and the presence of guanine nucleotides. The high affinity state corresponds with HR* and HR*G, whereas the low affinity state corresponds with the HR form, respectively. When guanine nucleotides are absent, the high affinity state is long lived, since HR*G cannot dissociate.[4,12,16,17] At physiological GTP concentrations, the trimeric $G_{\alpha\beta\gamma}$ form is ephemeral (*vide supra*) and after dissociation of the ternary (HR*G) complex, the receptor plus its agonist are free to catalyze the activation of another G protein (Fig. 5.1). This process enables signal amplification which has been demonstrated for a.o. phototransduction,[18] but also accounts for the simultaneous activation of the same G protein pool by different types of receptor molecules.

Samama et al[12] further link the apparent affinity of an agonist to its efficacy, not only in the G protein-coupled state of the receptor (HR*G) but also in its uncoupled state (HR*). The authors mention several receptor mutants with an increased tendency to form the active R* state. Agonist-binding behavior and spontaneous activation is explained by assuming that these mutants relax spontaneously to the active R* state.

Within the ETCM, the capability of an agonist to induce the formation of the ternary complex depends on two factors: (i) its ability to facilitate the transition from R to R* and (ii) its ability to stabilize the ternary complex HR*G. If the transition from R to R* is facilitated via e.g. mutations,[12] the apparent affinity of a ligand for its receptor is enhanced to a degree related to its efficacy both in the absence and presence of G protein. Basal (ligand-independent) activity stems from R*G; thus, when R* is formed more easily via e.g. mutations, basal activity also increases. There is indirect evidence that mutations[12] might also induce a higher affinity of R* for G. However, there is no definite proof, since it is not certain whether or not there is an amplification on the level of signal transmission from R*G·GTP to effectors such as adenylate cyclase.

Antagonists and Inverse Agonists Within the ETCM

Evidence exists that at least some antagonists display a high and low affinity pattern.[19] This pattern is, however, exactly opposite to the one displayed by agonists; in other words, antagonists bind more tightly to R and more loosely to R* (ref. 19, and refs. *loc. cit.*). The ETCM (also called two-state model,[13] cf. R and R*) is able to explain high and low affinity patterns observed for some antagonists, but also lays the basis for the reverse intrinsic activity as described by Schütz and Freissmuth.[19] These authors discriminate between so-called 'neutral' antagonists having 'null' intrinsic activity, and 'negative' antagonists with 'reverse' intrinsic activity. 'Negative' antagonists are also called *inverse* agonists. Inverse agonists lower basal activity, whereas 'neutral' antagonists do not affect this activity.

For examples of a constitutively active receptor as a disease-causing mechanism and possible therapeutic applications of inverse agonists to treat the kind of diseases connected to this pathology, the reader is referred to Milligan et al[20] and Parma et al.[21] Therapy with inverse agonists can, however, also have some disadvantages, because inverse agonism has been suggested to be a mechanistic basis for receptor upregulation.[22] In the case of the histamine H_2 receptor, the receptor density of a cell increases after prolonged incubation with inverse agonists, such as cimetidine, a drug widely used in the treatment of gastric ulcers.[22] An increase in receptor density leads to an increase in basal activity. The phenomenon of receptor upregulation might explain that long-term exposure to clinically used histamine H_2 receptor (negative) antagonists results in increased sensitization of the H_2 receptor,[23] increased intragastric hyperacidity[24] and loss of antisecretory effect.[25] On the other hand, treatment with neutral H_2 receptor antagonists would under normal conditions not cause major changes in receptor densities; concomitant changes in basal activity (measured as cAMP levels) after prolonged incubation with the neutral antagonist burimamide are indeed not observed.[22] These data imply a careful reconsideration of the pharmacological activity (neutral vs. negative) of clinically used GPCR antagonists. In some instances neutral antagonists may be favored. Abrupt withdrawal of drug therapy involving any receptor blockade by inverse agonists might result in hyperactivity caused by a certain receptor upregulation.

The Energetics of the ETCM

In the absence of agonists, basal activity can be observed. This implies that a part of the receptor population must be present in the R* state. From the Boltzmann distribution rule the energy difference between the R and R* state can be calculated once the [R]/[R*] ratio is known. G coupling to R* will stabilize R* with respect to R. Also coupling of H to R* or to R*G will cause a stabilization of the R* state (see also refs. 12,13,19,20). The mutation studies by Samama et al[12] on the β_2 adrenoceptor yielded a constitutively active receptor in the absence of agonist. These mutations must therefore have (partially) altered the energy difference between R and R*.

Certain antagonists bind more tightly to R than to R* (Schütz and Freissmuth[19]). From this it is inferred that inverse agonists (i.e., negative antagonists) can lower basal activity by stabilizing R (at the expense of R*) and therefore induce net uncoupling of G proteins (R does not couple to G). Hence, neutral antagonists are not expected to alter the energy difference between R and R*.

Conformational Changes Within the ETCM

As we have discussed, coupling of G proteins to a GPCR seems to be mediated by the intracellular loops of a GPCR (e.g., refs. 2,5,26-31, and refs. *loc. cit.*). Samama et al[12] and Jung et al[29] suggest that within the inactive conformation of the receptor (R state) these intracellular regions are shielded from interaction with the G protein. Thus according to this concept, a conformational change in the GPCR is responsible for its activation (R => R*). Only after activation of the receptor (yielding R*), is an interaction established, as a result of structural changes.

Also Weinstein[32] (and refs. therein) points to conformational changes as the origin for the activation of the G protein by a GPCR. Theoretical studies were performed on a seven helix model of the 5-HT_2 receptor in order to study ligand-induced displacements of the C_α carbons of the protein backbone; intra- and extracellular loops were not included in the model. From these data an activation model was inferred in which agonists distort the protein from the resting state with the extent of distortion related to the known pharmacological efficacies; antagonists did not distort the protein. The agonist-induced conformational changes were most pronounced in helices V and VI at the cytosolic side, strongly suggesting that the third intracellular loop will be affected, leading to G protein activation. However, it should be noted that the above calculations were restricted solely to the transmembrane helices and that besides the neglect of the intra- and extracellular loops any role for the membrane was also omitted. The helices were allowed to move freely in a medium with a dielectric permitivity of 4. It is important to mention that the above model does not address inverse agonism (i.e., negative antagonism) or neutral antagonism explicitly, since all agonists are assumed to distort the protein, whereas all antagonists are assumed not to alter the protein.

Oliveira et al[33] assign a switch function to a highly conserved arginine residue in the third membrane-spanning helix. This arginine also plays a crucial role in Timms's proton pumping model.[34,35] The arginine is thought to be located close to both the polar receptor pocket and the cytosol. Therefore, according to Oliveira et al[33] this residue could be shielded in the R state, whereas it becomes exposed in the R* state.

Hausdorff et al[36] report a mutation (a seven amino acid deletion in the carboxyl-terminal region of the third intracellular loop) of the β_2 adrenergic receptor that impairs agonist activation of the effector system (adenylate cyclase) without affecting high affinity agonist binding. This means that this mutation does not influence G protein binding, since the amount of G protein binding is considered to be directly related to high affinity agonist binding (see also ref. 37). In other words, the formation of high affinity ternary complexes is not sufficient to activate the G protein pool adequately. Therefore, it is concluded that the molecular determinants of ternary complex formation are not identical to those that transmit the agonist-induced stimulatory signal to the G protein,[36] showing that the concept of one particular conformational change

(R => R*) to explain G protein activation is questionable. Also, mutants for rhodopsin have been reported that are able to bind the G protein in a light-dependent manner but lack the ability to produce the activated G_α*·GTP form.[38]

5.3 SUMMARY

Within the extended ternary complex or two-state model, GPCRs can adopt two states: the inactive R and the active R* state. Agonists (H) bind with low and high affinity to the R and R* state, respectively. Inverse agonists (also called negative antagonists) display the opposite pattern, i.e., high affinity for R and low affinity for R*. Neutral antagonists do not discriminate between the R and R* state. G proteins couple to R* only, thereby enhancing the proportion R*. Agonists enhance the amount of R* and thus induce G protein coupling, whereas inverse agonists favor R, resulting in net uncoupling of the G protein.

If the R* state is considered as the state in which the receptor is coupled to the appropriate G protein, the species HR* cannot exist (Shankley, personal communication). Leeb-Lundberg et al[37] claim, however, to provide the experimental evidence for the existence of both the HR* and the HR*G state from the fact that high affinity agonist binding was still present at the uncoupled state of the B_2 bradykinin receptor: when a non-hydrolyzable GTP analogue was added, preventing coupling of G protein to the receptor, a considerable amount of high affinity binding could still be observed.[14] Furthermore, Strader and co-workers[15] even claim to have mutated a GPCR in such a way that it has become unable to couple to the G protein, yet in combination with enhanced agonist affinity. This raises the question whether the R* form in the ETCM combinations HR*, R*G and HR*G really is one and the same form. In addition, Hausdorff et al[36] have proven that the rate of G protein activation is not solely dependent on the level of high affinity binding. Therefore, the molecular determinants of G protein coupling and activation seem to be distinct. Explaining the amount of G protein activation by only a straightforward Boltzmannian relationship between high affinity binding of agonists and the amount of R*G present might thus be an oversimplification.

REFERENCES

1. Birnbaumer L, Mattera R, Yatani A et al. Recent advances in the understanding of multiple roles of G proteins in coupling of receptors to ionic channels and other effectors. In: Moss J, Vaughan M, eds. ADP-ribosylating toxins and G proteins, Insights into signal transduction. Washington DC: Am Soc Microbiol, 1990:225-266.
2. Birnbaumer L. G proteins and the modulation of potassium channels. In: Weston AH, Hamilton ThC, eds. Potassium channel modulators. Oxford: Blackwell Scientific Publications, 1993:44-75.
3. Birnbaumer L, Birnbaumer M. Signal transduction by G proteins: 1994 edition. J Receptor & Signal Transduction Research 1995; 15:213-252.
4. Chabre M, Antonny B, Vuong TM. The transducin cycle in the phototransduction cascade. NATO ASI Series, Series H 1991; 52:207-220.
5. Conklin BR, Bourne HR. Structural elements of G_α subunits that interact with $G_{\beta\gamma}$, receptors and effectors. Cell 1993; 73:631-641.
6. Gilman AG. G proteins: Transducers of receptor-generated signals. Annu Rev Biochem 1987: 56:615-649.
7. Gilman AG. G-Proteine und die Regulation der Adenylat-Cyclase (Nobel-Vortrag). Angew Chem 1995; 107:1533-1548.
8. Rodbell M. Signaltransduktion: Die Entwicklung einer Theorie (Nobel-Vortrag). Angew Chem 1995; 107:1549-1558.
9. Stryer L. Cyclic GMP cascade of vision. Ann Rev Neurosci 1986; 9:87-119.
10. Costa T, Ogino Y, Munson PJ et al. Drug efficacy at guanine nucleotide-binding regulatory protein-linked receptors: Thermodynamic interpretation of negative antagonism and receptor activity in the absence of ligand. Mol Pharmacol 1992; 41:549-560.
11. DeLean A, Stadel JM, Lefkowitz RJ. A ternary complex model explains the agonist-specific binding properties of the adenylate cyclase-coupled β-adrenergic receptor. J Biol Chem 1980; 255:7108-7117.
12. Samama P, Cotecchia S, Costa T et al. A mutation-induced activated state of the β_2 adrenergic receptor. Extending the ternary complex model. J Biol Chem 1993; 268:4625-4636.
13. Leff P. The two-state model of receptor activation. Trends Pharmacol Sci 1995; 16:89-97.
14. Leeb-Lundberg LMF, Mathis SA, Herzig MCS. Antagonists of bradykinin that stabilize a G protein-uncoupled state of the B2 receptor act as inverse agonists in rat myometrial cells. J Biol Chem 1994; 269:25970-25973.
15. Strader CD, Dixon RAF, Cheung AH et al. Mutations that uncouple the beta-adrenergic receptor from G_s and increase agonists affinity. J Biol Chem 1987; 262:16439-16443.

16. Vuong TM, Chabre M, Stryer L. Millisecond activation of transducin in the cyclic nucleotide cascade of vision. Nature 1984; 311:659-661.
17. Vuong TM, Chabre M. Subsecond deactivation of transducin by endogenous GTP hydrolysis. Nature 1990; 346:71-74.
18. Fung BK-K, Hurley JB, Stryer L. Flow of information in the light-triggered cyclic nucleotide cascade of vision. Proc Natl Acad Sci USA 1981; 78:152-156.
19. Schütz W, Freissmuth M. Reverse intrinsic activity of antagonists on G-protein-coupled receptors. Trends Pharmacol Sci 1992; 13:376-379.
20. Milligan G, Bond RA, Lee M. Inverse agonism: pharmacological curiosity or potential therapeutic strategy? Trends Pharmacol Sci 1995; 16:10-13.
21. Parma J, Duprez L, Van Sande J et al. Constitutive active receptors as a disease-causing mechanism. Mol Cell Endocrinol 1994; 100:159-162.
22. Smit MJ, Timmerman H, Alewijnse AE et al. Inverse agonism of histamine H_2 antagonists leads to antagonist-induced upregulation of the histamine H_2 receptor. Proc Natl Acad Sci USA 1996; 93:6802-6807.
23. Coruzzi G, Bertaccini G. Increased parietal cell sensitivity after chronic treatment with ranitidine in the conscious cat. Agents and Actions 1989; 28:215-217
24. Nwokolo CU, Smith JTL, Sawyer AM et al. Rebound intragastric hyperacidity after abrupt withdrawal of histamine H_2 receptor blockade. Gutt 1991; 32:1455-1460.
25. Merki HS, Wilder-Smith CH. Do continuous infusions of omeprazole and ranitidine retain their effect with prolonged dosing? Gastroenterology 1994; 106:60-64.
26. Collins S. Molecular structure of G protein-coupled receptors and regulation of their expression. Drug News & Perspectives 1993; 6:480-487.
27. Donnelly D, Findlay JBC, Blundell TL. The evolution and structure of aminergic G protein-coupled receptors. Receptors and Channels 1994; 2:61-78.
28. Hibert MF, Trumpp-Kallmeyer S, Hoflack J et al. This is not a G protein-coupled receptor. Trends Pharmacol Sci 1993; 14:7-12.
29. Jung H, Windhaber R, Palm D et al. NMR and circular dichroism studies of synthetic peptides derived from the third intracellular loop of the β adrenoceptor. FEBS Lett 1995; 358:133-136.
30. Lee NH, Kerlavage AR. Molecular biology of G protein-coupled receptors. Drug News & Perspectives 1993; 6:488-497.

31. Oliveira L, Paiva ACM, Vriend G. A common motif in G protein-coupled seven transmembrane helix receptors. J Comp-Aided Mol Design 1993; 7:649-658.
32. Weinstein H. Computational simulations of molecular structure, dynamics and signal transduction in biological systems: Mechanistic implications for ecological physical chemistry. In: Bonati L, Cosentino U, Lasagni M et al, eds. Trends in ecological physical chemistry, Proceedings of the 2nd international workshop on ecological physical chemistry. Amsterdam: Elsevier, 1993:1-16.
33. Oliveira L, Paiva ACM, Sander C et al. A common step for signal transduction in G protein-coupled receptors. Trends Pharmacol Sci 1994; 15:170-172.
34. Timms D, Wilkinson AJ, Kelly DR et al. Interactions of Tyr^{377} in a ligand-activation model of signal transmission through β_1 adrenoceptor α-helices. Int J Quant Chem: Quant Biol Symp 1992; 19:197-215.
35. Timms D, Wilkinson AJ, Kelly DR et al. Ligand-activated transmembrane proton transfer in β_1 adrenergic and m_2 muscarinic receptors. Receptors and Channels 1994; 2:107-119.
36. Hausdorff WP, Hnatowich M, O'Dowd BF et al. A mutation of the β_2 adrenergic receptor impairs agonist activation of adenylate cyclase without affecting high affinity agonist binding. J Biol Chem 1990; 265:1388-1393.
37. Burstein ES, Spalding TA, Bräuner-Osborne H et al. Constitutive activation of muscarinic receptors by the G protein G_q. FEBS Lett 1995; 363:261-263.
38. Ernst OP, Hofmann KP, Sakmar TP. Characterization of rhodopsin mutants that bind transducin but fail to induce GTP nucleotide uptake. J Biol Chem 1995; 270:10580-10586.

CHAPTER 6

Nucleotide Exchange Reactions and G Protein Activation

6.1 SURVEY OF EXPERIMENTAL EVIDENCE FOR AN EXCHANGE REACTION

Any model or theory is valid only as long as it explains all experimental observations. A new or adapted model for G protein activation should thus be introduced only when there are particular experimental findings that do not correspond with the established theory. Furthermore, this new model must account for all findings which can also be explained by the currently widely accepted exchange theory, including data concerning nucleotide exchange reactions which are outlined below. These current models (e.g., Fig. 5.1) assume GTP binding to HR*G or R*G to occur immediately after pre-bound GDP release. Since the release of pre-bound GDP from the G protein is thought to be the rate limiting step in the activation process, the role of the agonist and R* is suggested to be a catalytic one, increasing the rate of GDP dissociation.

Fung and Stryer[1] were among the first authors to consider such a catalytic cycle, in which the GTP for GDP exchange plays a crucial role. After incubation in the dark of rod outer segment (ROS) membranes containing high concentrations of rhodopsin, GDP (and not GTP) is incorporated and becomes tightly bound. Nearly all GDP remains bound in the dark to the G protein (*in casu* transducin, T-r). Photolysis is thought to induce release of GDP, which can be markedly enhanced by the presence of GTP,

GDP itself or non-hydrolyzable analogues of GTP such as p[NH]ppG or GTPγS. The kinetics of p[NH]ppG binding were compared to the kinetics of GDP release and it was concluded that p[NH]ppG exchanges for bound GDP after photolysis.[1] Moreover, it was found that the exchange process is highly amplified, i.e., one photolyzed rhodopsin molecule catalyzes the binding of up to 500 p[NH]ppG molecules.[1,2]

Fung and Stryer[1] suggested that GTP for GDP exchange can be established in two different ways: (1) R*·G·GDP releases GDP and subsequently binds GTP, (2) GTP binds to a second nucleotide binding site, which releases GDP from the first site. In both cases G_{α}*·GTP dissociates and R* can bind another G·GDP (*in casu* T-$r_{\alpha\beta\gamma}$·GDP). This recycling does not require GTP hydrolysis, since a high degree of amplification is found for p[NH]ppG (*vide supra*). Current models are based on the first mentioned possibility (cf. Fig. 5.1).

In the dark, transducin is almost entirely in the GDP state, although the formation of G_{α}*·GTP (i.e., T-r_{α}*·GTP) is thermodynamically feasible (Eq. 6.1). In the dark, the barrier for the GTP/GDP exchange and subsequent G_{α}*·GTP dissociation (Eq. 6.1) is probably too high.[1,3]

$$\text{T-r}_{\alpha\beta\gamma}\cdot\text{GDP} + \text{GTP} \rightleftharpoons \text{T-r}_{\alpha}\cdot\text{GTP} + \text{T-r}_{\beta\gamma} + \text{GDP} \qquad (6.1)$$

It has been inferred from hydrolysis-resistant analogues such as GTPγS that the equilibrium constant for this reaction is larger than 100 M.[3,4] Mg^{2+} shifts the equilibrium to the right. For a detailed review concerning the multiple effects of different Mg^{2+} concentrations on G protein activation, the reader is referred to Gilman.[5]

Therefore, the role of the activated receptor R* is suggested to lower the barrier of G protein activation. Thus, the level of T-r_{α}·GTP (G_{α}*·GTP) is primarily determined by the concentration of R*. G_{α}*·GTP is restored to the dark state by intrinsic GTPase activity. Both the formation of G_{α}*·GTP and its subsequent hydrolysis have been suggested to be thermodynamically favorable. In the dark state and at physiological GTP and GDP levels (both about 2 mM, ref. 6), G_{α}*·GTP is formed within milliseconds (*vide infra*).[7]

6.2 CAN THE CURRENT MECHANISM FOR G PROTEIN ACTIVATION BE FALSIFIED?

EXPERIMENTS THAT CHALLENGE THE ESTABLISHED MODEL

There are two literature findings in particular that do not comply with the current model of G protein activation as described in the previous subsection (Fig. 5.1). The *first* finding stems from experiments reported by Robinson and Hagins,[6] who call it the *usual large drop in GDP* observed within 4 s after photolysis of rhodopsin present in intact frog rod outer segments (ROSs; cf. Fig. 6.1). This observation clearly contradicts the proposed exchange mechanism of GDP for GTP in the current model in which GDP levels are likely to rise, and certainly not to drop in the period directly after activation (e.g. ref. 1). Furthermore, Robinson and Hagins[6] obtained strong evidence that in the first 4 GTP hydrolysis also occurred at a moderate level. This seemingly contradicts the observation that total GTP levels stay about constant in this time period.[6] Additionally, also in the period between 4 and 64 s after the flash, total GTP levels dropped, while GDP rose by similar amounts again indicating GTP hydrolysis.[6]

In intact frog ROSs, at least 100 GTP molecules are hydrolyzed for each photon absorbed by rhodopsin during the 20 s following a flash.[6] The concentration of rhodopsin in ROS is about 3 mM (ref. 3), while approximately 1% of the total photopigment is bleached in the photolysis experiments reported by Robinson and Hagins.[6] Since an amplification factor of more than 100 for R* activating G is claimed,[3,6] a huge GDP release is to be expected in case of an exchange mechanism (mM range) instead of the reported GDP drop.[6]

ROSs contain guanyl cyclases catalyzing the conversion of GTP into cGMP and cyclic nucleotide phosphodiesterases catalyzing the conversion of cGMP into GMP. The catalytic activity of these enzymes affect guanine nucleotide levels as well and might therefore interfere with conclusions drawn from the aforementioned experiments in which GTP and GDP concentrations are directly related to G protein activation alone.[6] Although the catalysis by phosphodiesterase is highly efficient (turnover ~$800 s^{-1}$; ref. 1), total acid-extractable GMP levels are much lower than extractable

GDP and GTP levels in ROSs before and after activation.[6] Moreover, GDP appears to be the main product of GTP metabolism in a time period of 64 s after illumination.[6] Hence, GTP hydrolysis to GDP is the main metabolic pathway, unless GMP is very rapidly phosphorylated to GDP (for the latter phosphorylation no evidence has been reported). That GTP hydrolysis is indeed the main route is also supported by the observation that within the first 4 s after the flash labelled GDP increase is matched by an equal loss in tritiated GTP and within 64 s the loss in total GTP is matched by an equal increase in GDP.

Also in dark-adapted ROSs, total acid-extractable GMP levels (μM range) are much lower than GDP and GTP levels (mM range), so that the GMP substrate/product concentrations related to the highly efficient phosphodiesterase are relatively low compared to the substrate/product concentrations available to the less efficient G protein. It is also shown that cGMP directly opens sodium channels at μM concentrations.[3] Hence, the concentration effect of guanine nucleotides due to the G protein activation (substrate at mM concentration) is more pronounced than the effect of phosphodiesterase (product at μM concentration). Furthermore, if the total concentration of phosphodiesterase and/or guanyl cyclase is much lower than that of transducin, the effect on total guanosine levels of the last mentioned protein will, despite its lower turnover number, be again much larger. Robinson and Hagins[6] measured total acid-extractable nucleotide levels in ROSs, and we have here referred to these levels throughout. Since the mass balance of guanine nucleotides in these experiments is not complete (cf. Fig. 6.1), a significant part of the guanine nucleotides must become acid-inextractable after activation (this phenomenon will be addressed later on). This does, however, not affect our remark that, immediately after rhodopsin activation, a large increase in (acid-extractable) GDP would be expected if the GDP/GTP exchange mechanism were indeed valid.

The *second* finding stems from phosphorylation reactions reported for several types of G proteins.[8-12] A phosphate transfer is observed which proceeds via the G_β unit of G proteins. A histidine residue on the G_β unit becomes phosphorylated and subsequently dephosphorylated resulting in an overall phosphorylation

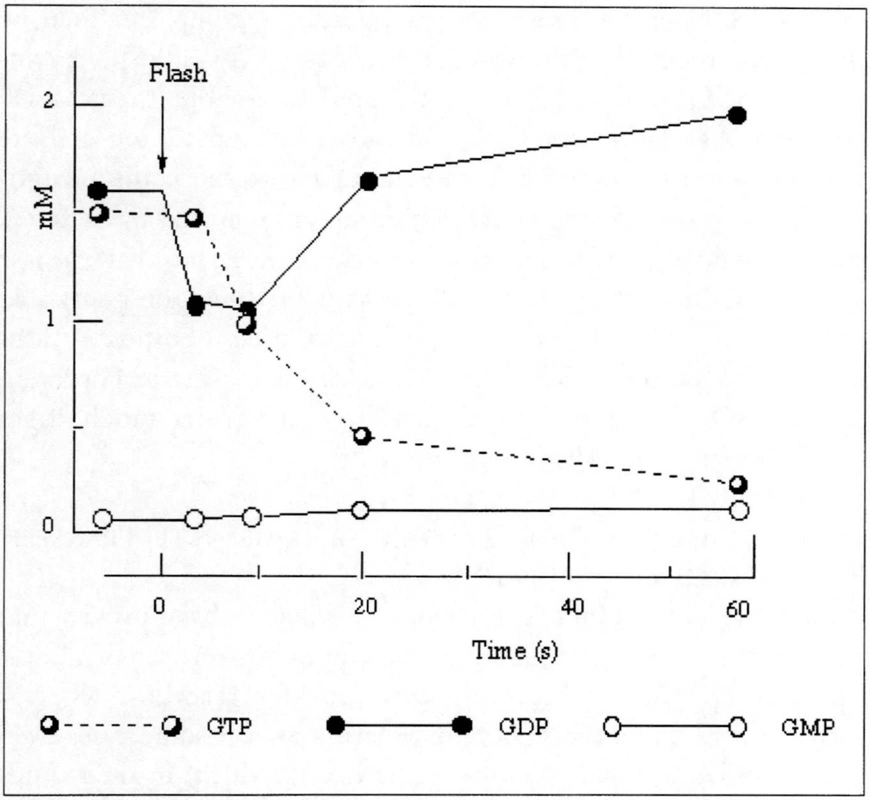

Fig. 6.1. Draft of (acid-)extractable GMP, GDP and GTP levels measured for intact isolated frog ROSs by Robinson and Hagins.[6] Time-scale starts at the beginning of a 2 ms flash of green light that bleached 1% of the rhodopsin chromophores. The depicted levels have been corrected for the small loss of all nucleotides that occurs in the dark during the 5-10 min course of the experiments, so there is no uncontrolled "dark" GTPase activity included. Because the main product of GTP metabolism in illuminated ROSs is GDP, it is assumed that GTP hydrolysis by G_α is the main biochemical pathway for GTP destruction, and that GMP does not become rapidly phosphorylated to GDP. Note that this GMP stems a.o. from cGMP which is formed from GTP by guanyl cyclase and which is subsequently converted into GMP by cyclic nucleotide phosphodiesterase, thereby releasing protons. Although phosphodiesterase has a much higher turnover number than G proteins, substrate/product concentrations for the diesterase are much lower than those for G proteins (GMP concentrations are in μM range, whereas both GDP and GTP levels are in the mM range).[3] GTP has a half time of about 10 s.[6]*

of GDP to GTP. This phosphorylation reaction is regulated by agonist-activated receptors. Wieland and co-workers[12] suggest that these phosphorylation reactions accelerate and amplify the receptor mediated signal. Kaldenberg-Stasch et al[8] report thiophosphorylation reactions with GTPγS, which are able to thiophosphorylate the G_β subunit of the G protein that couples to the GPCR fMet-Leu-Phe. The thiophosphate group can then be transferred to GDP yielding GTPγS (*in casu*, G·GTPγS). In contrast to GTP and GTPγS, p[NH]ppG is unable to establish transphosphorylation reactions and thus will bind to and activate G proteins via the classical nucleotide exchange mechanism. Similar results from a different research group are reported by Seifert[10], who concludes that T-r acts as a nucleoside diphosphate kinase (NDPK) and that the G_β subunit represents the (thio)phosphorylated enzyme intermediate (see chapter 4). Thus, a direct phosphate transfer reaction to the GDP bound to G_α in the trimeric G protein seems to be part of the G protein activation mechanism (Nederkoorn et al, 1995; *vide infra*).[13]

EVIDENCE AGAINST OR IN FAVOR OF A PHOSPHATE TRANSFER REACTION TO G_α SUBUNITS?

Very recently, Hohenegger et al[14] claimed to have proven that G_α subunits in platelet membranes are not the acceptor of the G_β subunit-dependent (thio)phosphate transfer reaction, whereas Kowluru and co-workers[15] report experiments (in some cases even rather similar to those of Hohenegger et al[14]) which, in their opinion, indicate that the G_α subunit does act as the (thio)phosphate acceptor. We discuss these conflicting findings by outlining the authors' experiments:[14,15]

(i) In accordance with aforementioned observations,[8-12] both Hohenegger et al[14] and Kowluru et al[15] conclude that guanine (and to a 10-fold lesser extent adenine) nucleotide-specific (thio)phosphorylation of G_β on a histidine residue indeed occurs in (G_β-containing) detergent extracts of membranes (of platelets[14] and pancreatic β cells,[15] respectively). Characteristic features of the (thio)phosphorylation of G_β subunits include the following findings. In native platelet membranes, maximum [^{35}S]labelling by [^{35}S]thiophosphorylation of G_β occurs within 1 min.[14] If G_β is labelled with [γ-^{32}P]ATP, more than 50% of the label is lost within 20 min,[15] indicating that G_β phosphorylation is ephemeral. Inclusion of GDP (10 μM) results in a marked acceleration in the dephosphorylation of G_β.[15] Thiophosphorylated

G_β subunits can be almost completely dephosphorylated by 100 μM GDP, whereas ADP and UDP have only a modest and a negligible effect, respectively.[14,15] G protein-coupled receptor-dependent regulation of the rate of (thio)phosphorylation of G_β is not observed in platelet membranes.[14] Hence, G_β phosphorylation does occur; it is, however, labile and GDP can act as the acceptor of the G_β (thio)phosphate.

(ii) To examine whether G_β subunits display auto(thio)phosphorylation activity, Hohenegger et al[14] and Kowluru et al[15] incubated both purified G_β subunits and $G_{\beta\gamma}$ dimers with [^{35}S]GTPγS or [γ-^{32}P]ATP. Under these conditions, (thio) phosphorylation of the G_β has not been found.[14,15] Subsequently, it was investigated whether or not purified G_α could induce the G_β (thio)phosphorylation. To this end, both purified G_β subunits and $G_{\beta\gamma}$ dimers were incubated with [^{35}S]GTPγS or [γ-^{32}P]ATP in the presence of purified G_α subunits. Again, (thio)phosphorylation of G_β was absent.[14,15] Hence, purified $G_{\beta(\gamma)}$ does not seem to be involved in an auto(thio)phosphorylation reaction and purified G_α subunits do not act as (thio)phosphate donors for $G_{\beta(\gamma)}$ (i.e., the trimeric G protein cannot be auto(thio)phosphorylated). In view of the findings mentioned in (i), Kowluru et al[15] suggest that other proteins than the G_α or G_β act as a source of the (thio)phosphate (*vide infra*, in viii).

(iii) To investigate whether (thio)phosphorylated G_β subunits can (thio)phosphorylate nucleotides bound to proteins, Hohenegger and co-workers[14] performed experiments in which membranes were pretreated with $NaIO_4$. Such a pretreatment causes the oxidation of 2'- and 3'-ribosyl hydroxyl groups of nucleotides to aldehydes, which, after reduction by $NaCNBH_3$ and $NaBH_4$, results in a covalent attachment of the nucleotide via a Schiff base to the ε-amino group of lysine residues. Following the pretreatment with $NaIO_4$, $NaCNBH_3$ and $NaBH_4$, [^{35}S]GTPγS was added to the platelet membranes, which resulted in the incorporation of radioactivity in many proteins—in particular in a 45-47 kDa protein—with a concomitant decrease in G_β radioactive labelling. Also, addition of [^{35}S]thiophosphate-labelled G_β to sequentially oxidized and reduced platelet membranes resulted in a similar radioactive labelling of proteins, mainly of a 45 kDa protein. This means that the G_β can donate its (thio)phosphate to nucleotides which are pre-bound to (acceptor) proteins.[14]

(iv) By covalently attaching the 2'-3'-dialdehyde GTPγS analogue (oGTPγS) to the appropriate proteins (a.o. G_α proteins), Hohenegger et al[14] could abolish the [^{35}S]thiophosphate donation by G_β. Since under these conditions the nucleotide binding site is occupied by covalently bound (cold) oGTPγS (unable to accept another (thio)phosphate), the result of this experiment indicates that a (thio)phosphate transfer reaction from G_β to a nucleotide bound at a protein does indeed occur. For the G protein, the difference with the experiments mentioned in (iii) lies in the fact that subsequent oxidation and reduction with $NaIO_4$, $NaCNBH_3$ and $NaBH_4$ mainly results in the covalent attachment of GDP to G_α, whereas here (iv) GTPγS is irreversibly attached to G_α.

(v) When 0.1 μM [^{35}S]GTPγS is added first, with subsequent oxidation and reduction of the platelet membranes, not only thiophosphorylated but also proteins that bind [^{35}S]GTPγS should become labelled.[14] Remarkably, the aforementioned (in iii) most prominent thiophosphate acceptor of 45 kDa lacks radioactive labelling under these conditions.[14] Therefore, the sequence of these reactions (first addition of [^{35}S]GTPγS, then oxidation/reduction) must have barred the (thio)phosphotransfer to the acceptor. In our opinion, this illustrates the pitfalls of such harsh treatment of the proteins present in the platelet membranes, and that conclusions drawn from these kinds of experiments might be ambiguous. The difference between (iii) and

(v) is that in (iii) G_β subunits are part of trimeric G proteins during oxidation/reduction, whereas in (v) 0.1 μM GTPγS is added first, which results in the activation of G proteins and the G_β subunits are thus (at least partially) in the $G_{\beta\gamma}$ dimer form during the oxidation and reduction. It is noteworthy that Ras proteins, especially, absorbed radioactivity under the latter experimental circumstances.

(vi) Sequential oxidation and reduction of platelet membranes abolish adenylate cyclase activity, which impairs the monitoring of the functional effect of phosphate transfer reactions.[14] However, after a brief pre-incubation with 2',3'-dialdehyde GTP (oGTP) followed by the addition of GTPγS, adenylate cyclase inhibition can still be monitored, whereas α_2 adrenergic receptor stimulation even increases adenylate cyclase inhibition under these circumstances (N.B. α_2 adrenergic receptors mediate G_i protein activation). When, after a brief incubation with oGTP, p[NH]ppG is added (p[NH]ppG is not able to participate in a phosphotransfer reaction), adenylate cyclase activity is not inhibited. These findings led Hohenegger et al[14] to conclude that a receptor-stimulated phosphate transfer reaction is indeed present (N.B. This phosphotransfer is blocked by 1mM UDP). However, in the same article[14] the authors use the fact that prolonged incubation with oGTP—eventually resulting in the formation of G_i·oGDP thereby fully inactivating G_i (but probably also affecting adenylate cyclase activity more directly[16])—with subsequent addition of GTPγS neither inhibited nor stimulated adenylate cyclase, as an argument against a possible (thio)phosphate transfer reaction occurring directly on GDP bound to G_α. Thus, on the one hand a brief incubation with oGTP is necessary to preserve adenylate cyclase activity, but, on the other hand the absence of alterations in adenylate cyclase activity after prolonged incubation with oGTP is used as an argument against a possible (thio)phosphate donation at G_α.[14] Since it cannot be excluded that prolonged exposure to oGTP affects the adenylate cyclase protein directly, we feel that any conclusions drawn from experiments in which the alterations in cAMP levels are monitored after a long-term incubation with oGTP should be treated with caution.

(vii) $G_{s\alpha}$ and $G_{i\alpha 2}$ can be immunoprecipitated with the appropriate antisera. If platelet membranes are oxidized, reduced and subsequently incubated with [^{35}S]GTPγS (*vide supra*, in iii), radioactively labelled $G_{s\alpha}$ or $G_{i\alpha}$ is not found in the immunoprecipitation.[14] This implies that neither $G_{s\alpha}$ nor $G_{i\alpha}$ function as a thiophosphate acceptor, unless the oxidation or reduction reactions distort the thiophos–photransfer to G_α or the antiserum binding. Hohenegger et al[14] claim that antiserum binding is not altered by the oxidation and reduction reactions. However, Kowluru et al[15] indicate that the immunoprecipitation procedure itself can indeed affect the phosphorylation state of a protein: when an antiserum to the G_β subunit is used, significant losses of G_β subunit-associated radioactivity, which is directly related to the G_β phosphorylation state, are observed during the immunoprecipitation process. This phenomenon seems to be common to nucleotide diphosphate kinases (NDPKs; ref. 15 and refs. *loc. cit.*). Therefore, antiserum binding could have prevented the thiophosphate transfer towards the GDP at G_α in Hohenegger and co-workers' experiments by an actual dethiophosphorylation of the G_β. If, on the other hand, [^{35}S]GTPγS is added first followed by oxidation and reduction reactions, immunoprecipitated radioactively labelled $G_{s\alpha}$ or $G_{i\alpha}$ is found, indicating that a nucleotide exchange reaction of [^{35}S]GTPγS for pre-bound GDP can take place under these conditions.[14]

(viii) UDP is known to inhibit the activity of NDPKs (ref. 15 and refs. *loc. cit.*). UDP itself is unable to accept the (thio)phosphate of G_β,[14] yet it inhibits the phosphorylation of G_β.[15] G_β (alone, as part of the $G_{\beta\gamma}$ dimer, or as part of the trimeric $G_{\alpha\beta\gamma}$ protein) is proven not to be involved in autophosphorylation[14,15] (*vide supra*, in ii), and hence by definition, neither the G_β itself nor the trimeric G protein itself are NDPKs. Because the addition of membranes is required for G_β phosphorylation and because GTP and ATP can be used as phosphate donors for G_β, a membrane-associated NDPK is suggested to be involved in the regulation of this G_β phosphorylation.[15] Furthermore, the phosphorylation of G_β is stimulated by spermine and inhibited by heparin, which points to an involvement of casein kinase II-like enzymes.[15] These findings conflict with Seifert's conclusion[10] (see previous section) that T-r itself acts as a NDPK.

(ix) Besides free GDP (and to a lesser extent ADP[14] or IDP[15]) G_α·GDP also is found to serve as a receiver of the phosphate from the G_β subunit. Hohenegger et al[14] conclude that G_β phosphorylates only free GDP to free GTP and that subsequently the GTP formed in this way could participate in an exchange reaction at G_α. In other words, these authors[14] stick to the exchange mechanism for G protein activation. In contrast, Kowluru and others[15] developed two different approaches to demonstrate a phospho-relay from G_β to G_α, based on the formation of the hydrolysis-resistant GTP analogues, GTPγS and GTPβS, respectively: (a) If G_β is prelabelled with [γ-^{32}P]ATP and GDPβS is liganded to G_α, the formation of GT^{32}PβS is observed; (b) if G_β is prelabelled with ATPγ^{35}S and GDP is liganded to G_α, GTPγ^{35}S is formed. The addition of purified G_α liganded with GDPβS or GDP to (a) and (b), respectively, yields a greater than two-fold increase in the formation of nucleoside triphosphates (NTPs; i.e. GT^{32}PβS [a] and GTPγ^{35}S [b]).[15] Kowluru et al[15] conclude from these data the emergence of the active G_α^* form of G_α upon incubation of phosphorylated G_β with

G_α·GDP. Moreover, as a negative control, heat-inactivated G_α·GDP can be used and these heat-inactivated G_α subunits are found to be unable to dephosphorylate G_β.[15] Hence, a phosphotransfer from G_β towards G_α is assumed for which the native conformation of G_α complexed with GDP is a prerequisite.[15]

(x) The intact trimeric conformation of the G protein is probably required for G_β phosphorylation.[15] This is shown by the fact that pre-treatment with GTPγS or the GTP transition state analogue GDP·AlF_4^- reduced G_β labelling with [γ-^{32}P]ATP.[15] It is noted that GTPγS might inhibit ^{32}P labelling of G_β (with the ^{32}P stemming from [γ-^{32}P]ATP) by thiophosphorylating the G_β, but the transition state analogue cannot interfere at this level.

In summary, the purified G_α does not function as (thio)phosphate donor for purified $G_{\beta\gamma}$. The G_β very likely becomes (thio)phosphorylated by a membrane-associated kinase, and hence G_β is itself not an NDPK. In platelets, agonist influences (including activation of the A_2 adenosine receptor [G_s-coupled], α_2-adrenergic receptor [G_i-coupled] and thrombin receptor [G_i-, G_q- and $G_{12/13}$-coupled]) on the phosphorylation state of G_β are absent, although a receptor-stimulated phosphotransfer step is found to be involved in effector regulation.[14] An agonist influence on the phosphorylation state of G_β can, however, be found in human leukemia (HL-60) membranes.[12] A phosphotransfer from G_β to the nucleotide bound at G_α is not found by Hohenegger et al,[14] whereas Wieland and others[8-12] have laid the basis for Kowluru and co-workers[15] to explicitly claim that there is enough experimental evidence for exactly such a transfer. As immunoprecipitation procedures themselves can interfere with the phosphorylation state of proteins,[15] it is hard to draw final conclusions from these kind of experiments alone. Furthermore, we conclude that the reduction/oxidation experiments reported by Hohenegger and co-workers[14] do not give a definite answer to the question of whether or not the GDP bound to G_α in the $G_{\alpha\beta\gamma}$ trimer does act as the acceptor of a (thio)phosphate group, since we cannot rule out that the addition of $NaIO_4$, $NaCNBH_3$ and $NaBH_4$ blocks a possible (thio)phosphotransfer to the nucleotide pre-bound at G_α. In this respect, remember that heat-inactivated G_α·GDP cannot serve as a phosphate acceptor, whereas the native G_α·GDP can.[15] In this context it is also worthy of note that thiophosphotransfers are seen to be dependent on the sequence of addition of reactants: if $NaIO_4$, $NaCNBH_3$ and $NaBH_4$ are added, followed by GTPγS, thiophosphorylation of a 45 kDa protein is found, but, remarkably, when GTPγS is added first followed by $NaIO_4$, $NaCNBH_3$ and $NaBH_4$ (which should lead to labelling of a greater pool of proteins; *vide supra*, in v), thiophosphorylation of a 45 kDa protein cannot be found.[14] Moreover, adenylate cyclase activity is indicated to be sensitive to the aforementioned oxidation and reduction reactions.[14] Hence, G protein functioning could have been altered as well. Since all evidence against a possible phosphate transfer from G_β towards G_α subunits is based on immunoprecipitation and reduction/oxidation experiments,[14] we conclude that, in view of the indicated pitfalls for these types of experiments, the balance of evidence favors the possibility of such a phosphotransfer.

In platelet membrane preparations, the thiophosphorylation of G_β is seen to be independent of agonists of GPCRs, which could imply that the G protein is functionally decoupled from its receptor. Regarding the interaction of a GPCR to an appropriate G protein we note that proteases can prevent or alter this coupling (from the reported experimental procedures it is not clear whether protease inhibitors were indeed added at all stages[14]). In HL-60 membrane preparations[12] the functional GPCR-G protein coupling could have been preserved, since here agonist-dependence of the (thio)phosphorylated state of G_β is truly observed. Despite the fact that in the platelet membrane preparations high affinity agonist binding can be found,[17] this does not imply that the G protein is functionally coupled to its appropriate receptors, because the molecular determinants of G protein binding (associated with high affinity binding) and the actual G protein activation seem to be distinct (e.g., ref. 18; cf. chapter 5).

Inconsistencies Between Microcalorimetry and Guanine Nucleotide Levels

In addition to the aforementioned experimental findings that challenge the current model for G protein activation, other experiments have been performed, which were originally thought to be in support of the classical model, but which are actually more in line with our revised mechanism (cf. chapters 7 and 8). We refer in this respect to the conclusions drawn by Chabre and co-workers from microcalorimetric experiments on intact bovine ROSs.[19,20] Their conclusions (*vide infra*) are not in accordance with the nucleotide levels as measured by Robinson and Hagins[6] for frog ROSs (Fig. 6.1) and might contradict results from light scattering studies[7] as well.

Vuong et al[7] applied light scattering experiments on frog ROSs and observed a rapid increase in axial scattering, with a subsequent loss of scattering both in the axial as well as in the radial direction within the first millisecond after activation. The rapid axial increase in scattering probably arises from the release of transducin from the disk membrane surface into the inter-disk space enabling coupling to rhodopsin. The decrease in scattering seen in both the radial and axial signals has been associated with the dissociation of G_α*·GTP (due to the loss of soluble T-r_α from ROS segments). These experiments clearly demonstrate that the 'first round' of G_α*·GTP formation as presented in Eq. 6.1 proceeds on a *millisecond* scale: transducin in the dark state with pre-bound GDP couples to R* followed by G_α*·GTP dissociation. When T-r_α* recombines with the T-$r_{\beta\gamma}$ dimer, the trimer can again be activated and contributes to the second, third etc. rounds of activation. By adding hydroxylamine to rhodopsin (causing R* to disappear), reactivation of the trimer is prevented.

Chabre et al[19,20] conclude from microcalorimetric studies that GTP hydrolysis following G_α*·GTP dissociation must be finished on a *subsecond* scale. This conclusion was drawn from experiments on activated ROSs in which a characteristic heat pulse was observed within one second after activation. The heat release was ascribed to GTP hydrolysis of the first round[19,20] (cf. Eq. 6.1), suggesting that all transducin with pre-bound GDP becomes quickly activated and almost synchroneously deactivated by GTP hydrolysis. In the absence of hydroxylamine (a compound that is known to inactivate rhodopsin), this characteristic heat pulse is followed by a long tail, indicating more than one round of activation and also a slow turnover number for GTP hydrolysis.

Stryer[3] (current model; cf. Fig. 5.1) states that the vision excitation cycle for rhodopsin is driven by GTP hydrolysis and not by the energy of the absorbed photon. However, the phenomenon that the G_α* dissociation reaction (Eq. 6.1) is exergonic ($\Delta G << 0$; K_{eq} 100 M, *vide supra*) is not taken into account. Note that non-hydrolyzable GTP analogues also show large amplification, which clearly indicates the exergonic character of Eq. 6.1: in the latter case, no energy from hydrolysis is available for the vision excitation cycle, yet there is a large amplification. Moreover, the hydrolysis reaction proceeds more slowly (subsecond scale according to Vuong and Chabre[19]) than the preceding exchange reaction (huge amplification) for which Vuong et al[7] measured the release of transducin and subsequent G_α* dissociation from the disk membrane surface into the inter-disk space on a millisecond scale. Since hydrolysis proceeds more slowly, and is even totally absent for non-hydrolyzable GTP analogues, the question is raised: from where does the energy for amplification on the millisecond scale in the current model originate?

The conclusions drawn from microcalorimetric studies[19,20] contradict the experimentally observed nucleotide levels as measured by Robinson and Hagins.[6] The latter authors observe a half-time for GTP hydrolysis in intact frog ROSs of about 10 s.[6] This clearly contradicts the conclusions drawn by Chabre et al[19,20] that G protein activation and subsequent GTP hydrolysis must be completed within the first second. We state that we can bring the results of the light scattering experiments, microcalorimetry and nucleotide level measurements on one line. For that purpose it is necessary to remember that not only GTP hydrolysis is thermodynamically favorable, but also G_α*·GTP dissociation (cf. Eq. 6.1 with K_{eq} >>1 M; for GTPγS K_{eq} is even larger than 100 M, and for p[NH]ppG amplification is also reported, indicating the exergonic character). The observation of the heat pulse at a subsecond time scale is then explained by assuming that the heat burst is a result of G_α*·GTP formation occurring on a millisecond basis, while the heat produced by the reaction is transferred relatively slowly via vibrations. Hence, the exergonic dissociation reaction proceeding on a millisecond scale (well within the 1-2 s rise time of the change in membrane potential[3]) gives rise to a heat pulse measured on a subsecond scale. GTP hydrolysis can then indeed occur on a larger time scale ($t_{1/2}$ of ~10s). This thesis could be tested experimentally: in the absence of GTP, but in the presence of a non-hydrolyzable analogue, the same amount of heat release as for GTP should be observed.

Chabre et al[20] raise the argument that there is a physiological need of the alleged subsecond GTP hydrolysis, because this hydrolysis should account for a quick deactivation process. However, Erickson et al[21] gathered evidence that GTP hydrolysis is not required for the deactivation of visual transduction, since deactivation processes are also observed for non-hydrolyzable GTP analogues. The deactivation of the second messenger system is suggested to occur via an inhibitory factor that is available even when GTP hydrolysis is blocked. Therefore, we conclude that there is not an absolute physiological necessity for a subsecond GTP hydrolysis process.

6.3 SUMMARY

Two experimental findings in particular, previously reported in the literature, do not comply with the current model of G protein activation as described in chapter 5 (Fig. 5.1): (i) guanine nucleotide levels in ROSs measured upon GPCR activation and (ii) (thio)phosphorylation reactions at the G protein. Although there are some questions unanswered regarding the precise nature of these (thio)phosphorylation reactions,[14] we conclude that the balance of evidence favors the option that the de(thio)phosphorylation of the G_β subunit is accompanied by concomitant (thio)phosphorylation of the pre-bound GDP at the G_α subunit, resulting in the activated G_α*·GTP.[13,15] The unaltered native state of G_α·GDP is a prerequisite for the phosphorylation of the pre-bound GDP to yield GTP.[15] All authors (e.g., refs. 8-15) do, however, agree on the fact that a (thio)phosphate transfer reaction is involved in effector regulation by the G protein.

Heat release upon rhodopsin activation in ROSs is measured on a subsecond time level and has been ascribed to GTP hydrolysis by G proteins. However, the half time of GTP in ROSs of 10s seems not to be in accordance with this subsecond time scale. Moreover, the amplification of one R* activating many G proteins has been suggested to be energetically fed by GTP hydrolysis.[3] There is, however, already a huge amplification of G protein activation on a millisecond scale. Therefore, we propose that the activation step of the G protein is exergonic itself. In other words, G protein activation by R* releases heat. Then it can also be explained why in the presence of non-hydrolyzable GTP analogues (not able to provide hydrolysis energy) amplification of G protein activation is found and why the heat release is measured on a subsecond time scale, whereas $t_{1/2}$ of GTP equals 10 s. Whether the dissociation of the G protein is indeed an exergonic reaction could be tested experimentally; in the presence of non-hydrolyzable GTP analogues, the same amount of heat release is expected as for GTP itself.

References

1. Fung BK-K, Hurley JB, Stryer L. Flow of information in the light-triggered cyclic nucleotide cascade of vision. Proc Natl Acad Sci USA 1981; 78:152-156.
2. Fung BK-K, Stryer L. Photolyzed rhodopsin catalyzes the exchange of GTP for bound GDP in retinal rod outer segments. Proc Natl Acad Sci USA 1980; 77:2500-2504.
3. Stryer L. Cyclic GMP cascade of vision. Ann Rev Neurosci 1986; 9:87-119.
4. Yamanaka G, Stryer, L. Interactions of transducin with two hydrolysis-resistant GTP analogues. Invest Ophthal Vis Sci (Suppl.) 1994; 25:157.
5. Gilman AG. G proteins: Transducers of receptor-generated signals. Annu Rev Biochem 1987; 56:615-649.
6. Robinson WE, Hagins WA. GTP hydrolysis in intact rod outer segments and the transmitter cycle in visual excitation. Nature 1979; 280:398-400.
7. Vuong TM, Chabre M, Stryer L. Millisecond activation of transducin in the cyclic nucleotide cascade of vision. Nature 1984; 311:659-661.
8. Kaldenberg-Stasch S, Baden M, Fesseler B et al. Receptor-stimulated guanine nucleotide-triphosphate binding to guanine nucleotide-binding regulatory proteins. Eur J Biochem 1994; 221:25-33.
9. Klinker JF. Transducin possesses NTPase activity. Naunyn-Schmiedeberg's Arch Pharmacol 1995; 351:R68.
10. Seifert R. Transducin is a nucleoside diphosphate kinase. Naunyn-Schmiedeberg's Arch Pharmacol 1995; 351:R68.
11. Wieland T, Kaldenberg-Stasch S, Fesseler B et al. Regulation of G-protein function by phosphorylation. Can J Physiol Pharmacol 1994; 72:S5.

12. Wieland T, Nürnberg B, Ulibarri I et al. Guanine nucleotide-specific phosphate transfer by guanine nucleotide-binding regulatory protein β-subunits. J Biol Chem 1993; 268:18111-18118.
13. Nederkoorn PHJ, Timmerman H, Donné-Op den Kelder GM. Does the ternary complex act as a secondary proton pump and a GTP synthase? Trends Pharmacol Sci 1995; 16:156-161.
14. Hohenegger M, Mitterauer T, Voss T et al. Thiophosphorylation of the G protein β subunit in human platelet membranes: evidence against a direct phosphate transfer reaction to G_α subunits. Mol Pharmacol 1996; 49:73-80.
15. Kowluru A, Seavey SE, Rhodes CJ et al. A novel regulatory mechanism for trimeric GTP-binding proteins in the membrane and secretory granule fractions of human and rodent β cells. Biochem J 1996; 313:97-107.
16. Nanoff C, Böhm S, Hohenegger M et al. 2',3'-Dialdehyde GTP as an irreversible G protein antagonist. J Biol Chem 1994; 269:31999-32007.
17. Hüttemann E, Ukena D, Lenschow V et al. R_a adenosine receptors in human platelets. Naunyn-Schmiedeberg's Arch Pharmacol 1984; 325:226-233.
18. Hausdorff WP, Hnatowich M, O'Dowd BF et al. A mutation of the β_2 adrenergic receptor impairs agonist activation of adenylate cyclase without affecting high affinity agonist binding. J Biol Chem 1990; 265:1388-1393.
19. Vuong TM, Chabre M. Subsecond deactivation of transducin by endogenous GTP hydrolysis. Nature 1990; 346:71-74.
20. Chabre M, Antonny B, Vuong TM. The transducin cycle in the phototransduction cascade. NATO ASI Series, Series H 1991; 52:207-220.
21. Erickson MA, Robinson P, Lisman J. Deactivation of visual transduction without guanosine triphosphate hydrolysis by G protein. Science 1992; 257:1255-1258.

CHAPTER 7

A New Mechanism for G Protein Activation

7.1 A SCHEMATIC VIEW OF A NEW MECHANISM FOR G PROTEIN ACTIVATION, PROTON-TRANSPORTING RECEPTORS AND GTP SYNTHESIS

We present a new model, which is mainly based on observations stemming from rod outer segments (ROSs), since in these segments transducin is largely abundant and the system is therefore ideal for studying G protein activation. Throughout, we assume that findings for one particular type of G protein present in a certain species and tissue can be used to infer the activation mechanism for G proteins in general. Within the new model, all experimental findings concerning exchange reactions are explained, as well as the observation that phosphorylation reactions regulate G protein functioning.[1-4] Also the seemingly contradictory results and interpretations of Vuong and co-workers[5-7] and the observed nucleotide levels reported by Robinson and Hagins[8] are brought into accordance within this model. Figure 7.1 summarizes our ideas regarding, in our opinion, a more satisfactory mechanism for G protein activation than is provided by the classic model; the new model is discussed step by step in the subsequent subsections.

Receptor Activation Coupling of the G Protein to the Receptor

The receptor R is assumed to be activated by either light (opsins) or hormone/neurotransmitter action yielding R*. Within the extended ternary complex model (ETCM), it is suggested that only

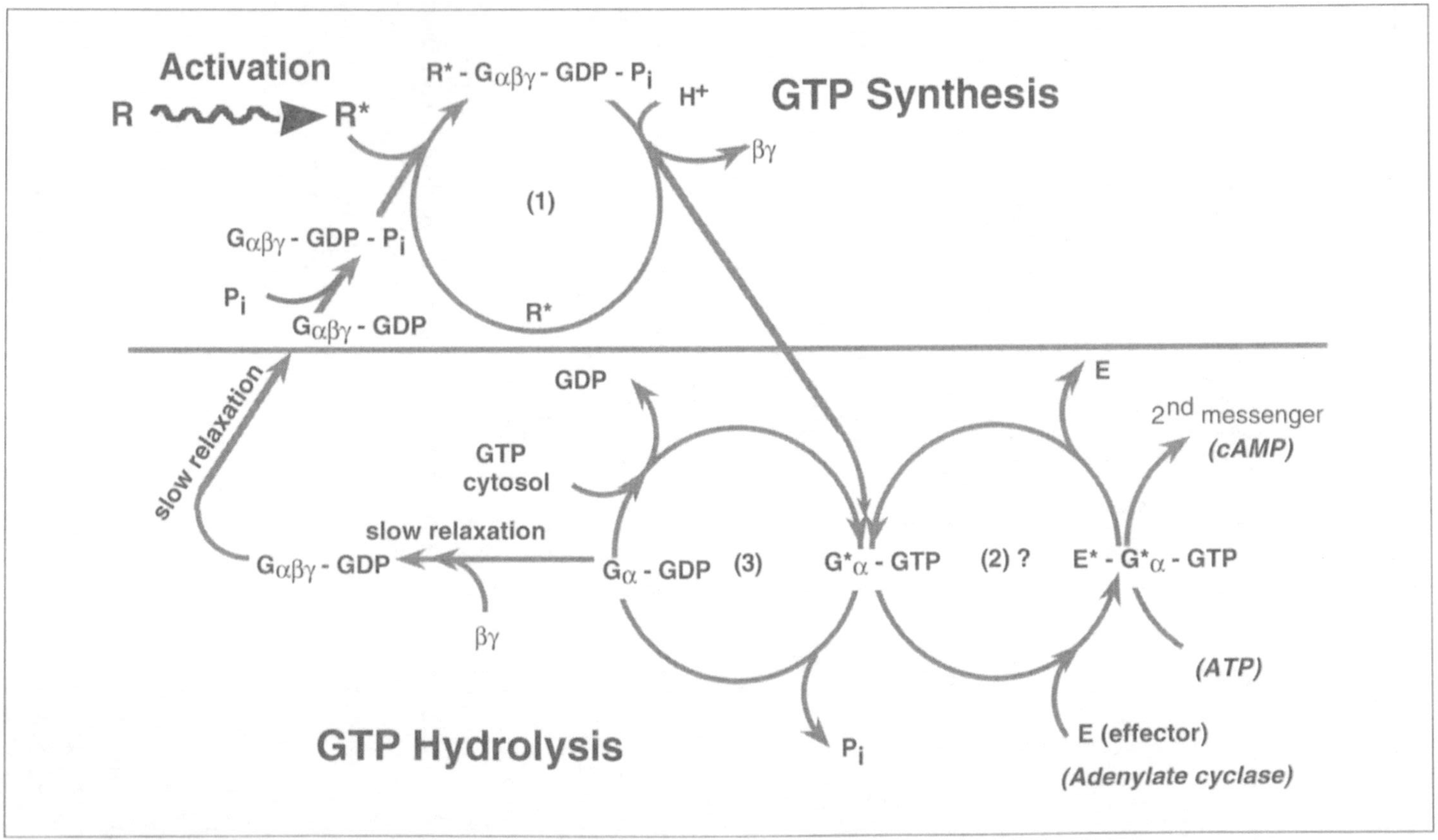

Fig. 7.1. New activation mechanism for G proteins. See text for further details. (Reprinted with permission from: Nederkoorn PHJ, Timmerman H, Donné-Op den Kelder GM. Does the ternary complex act as a secondary proton pump and a GTP synthase? Trends Pharmacol Sci 1995; 16:156-161.)

R* couples to G (cf. chapter 5 and ref. 9), or actually to a (multimeric) cluster of G proteins.[10] Also in our scheme, the first step in G protein activation is the coupling of the receptor to an appropriate (cluster of) G protein(s). Intracellular loops are considered to play a key role in the conversion from the R towards the R* state. As an example, the importance of the intracellular loops in G protein activation is confirmed by mutation studies on the hamster β_2 adrenergic receptor, which reveal that deleting residues 222-229 and amino acids close to the C-terminal of the third intracellular loop (258-270) results in a large reduction of agonist-dependent G protein stimulation (G_s).[11] In addition, the first and second intracellular loops and the C-terminal domain are reported to mediate G protein coupling and effector stimulation/inhibition (e.g., ref. 12 and refs. therein). In contrast, when residues 229-258 of the third cytoplasmic loop of the hamster β_2 adrenoceptor are deleted, no detectable effect on the ability of the receptor to stimulate adenylate cyclase is observed.[11] From mutation studies it is also known that G protein coupling to R* and/or the presence of agonistic high affinity binding does not automatically result in G protein activation.[11,13,14] The underlying molecular mechanisms are therefore suggested to be distinct (cf. chapter 5). In the next subsection, we describe such an activation mechanism, which clearly discriminates between plain G protein binding to the receptor and the actual activation process.

G Protein Activation, GTP Synthesis and Proton Pumping

The next step in our model (Fig. 7.1) is a GTP synthesis step and not the classical GDP/GTP exchange reaction. Considering the more open nucleotide binding cleft in G_α compared to the closed cleft in G_α^*,[15] the resulting synthesis mechanism can explain why GDP levels do not rise and GTP levels do not decrease when ROSs are activated (Fig. 6.1). Pre-bound GDP is assumed to be acid-extractable from (the more open) G_α structure, whereas—after GDP is converted into GTP, which in turn initiates G_α^* formation—the closed nucleotide binding site of G_α^* bars the (acidic) GTP extraction. These phenomena explain not only the fast decrease in GDP but also the apparent disappearance of total guanine nucleotides within the first 4 s after activation (cf. Fig. 6.1). The very fast G_α*·GTP formation (GDP levels decrease immediately after activation), together with the long half-time for GTP hydrolysis (~10s, ref. 8) combined with the fact that the nucleotide can possibly not be (acid-)extracted from G_α* (*vide infra*), explains that total (acid-)extractable GTP concentrations hardly change the first 4s after photolysis of rhodopsin.

At the GTP synthesis level, amplification can occur (level 1, Fig. 7.1). This amplification must occur on a millisecond time scale which is inferred from the very fast movement of transducin proteins to membrane-bound light-activated rhodopsin molecules.[7] Furthermore, we can now explain why GDP accelerates observed exchange reactions.[16] Because (pre-bound) GDP plus a phosphate (P_i) are consumed in the synthesis process, and GDP plays an important role in the activation process, resulting in dissociated G_α* subunits (which are suggested to be involved in the classical GDP/GTP exchange mechanism, *vide infra*). High GDP levels (more G proteins with pre-bound GDP) will therefore accelerate this process. The presence of multimeric clusters of G proteins[10] is confirmed by the first amplification level in Figure 7.1. Rodbell[10] suggests that one GPCR can activate a number of G proteins (all pre-coupled with GDP) which are grouped in such a cluster. Also the phosphorylation reactions reported in literature for G proteins, and acting at the GTP synthesis level 1, are in full agreement with our proposal, since it is found that these phosphorylation reactions result in an acceleration and amplification of the receptor-mediated signal.[1-4]

In strong analogy with ATP synthases, we propose that both the ternary complex HR*G and R*G can convert G_α into G_α* with a simultaneous change of GDP plus P_i into GTP (a phosphorylation reaction) with the aid of proton(s) which are chemical and/or vectorial (see Part I). Implicitly, we thus assume that the ternary complex acts not only as a GTP synthase, but also as a secondary proton pump consuming a protonmotive force generated by a primary pump in order to synthesize GTP. The P_i necessary for the synthesis reaction stems from the phosphorylated histidine residue at G_β. The introduction of P_i at G_α during the activation process prevents GDP from being released from the G_α

subunit by an activated R*, just like AlF_4^- can induce the transition state G_α*·GDP·AlF_4^-. The Gibbs energy required for the movement of (multimeric clusters of) transducins towards and along R* (kinetic energy) might also be delivered by the protonmotive force. The translocated protons are assumed to facilitate G_α*·GTP dissociation at physiological GTP levels, resulting in soluble G_α*·GTP and concomitant release of heat. We will address the proton pumping and GTP synthesis mechanism in more detail in chapter 8. Whether or not the GTP synthesis step is a prerequisite for G protein activation may be tested experimentally with [NH]ppG, since this nucleotide analague cannot participate in a synthesis reaction. Loading the G protein with pre-bound [NH]ppG in the absence of GDP should then inactivate the G protein. It is noted that GDPβS should not be considered as a negative control since Kowluru and co-workers[4] have demonstrated that this analagon can indeed be converted into GTPβS, leading to activated G_α*·GTPβS complexes (cf. chapter 6), which strongly supports the GTP synthesis hypothesis.

Effector Coupling

G_α*·GTP has a high affinity for its effector system. At this level, amplification might occur so that one G_α*·GTP activates more than one effector molecule. According to Samama et al[9] this amplification is not proven (this explains the question mark in Figure 7.1 at level 2).

Depending on the [GTP]/[GDP] and $[G_\alpha(*)]/[G_{\beta\gamma}]$ ratios, it can be thermodynamically favorable for one G_α* subunit to hydrolyze many GTP molecules (amplification level 3), thereby flipping between the active G_α* state to the inactive G_α state. During this hydrolysis process (3), G_α* can couple to effector (E) molecules. Effectors thus stabilize the G_α* state, explaining the ability of the effector system to speed up the GTPase activity of G_α* (refs. 17-18, and refs. *loc. cit.*). Furthermore, when G_α* lives longer, it can hydrolyze more GTP molecules (amplification level 3). The k_{cat} for the hydrolysis reaction is low, i.e., ~5 min^{-1} (e.g. refs. 17-18, and refs. *loc. cit.*). However, effector molecules are known to speed up the GTPase activity of G_α* up to 50 to 250 min^{-1}.[19] Given a half time of 10s in which 0.5 mM acid-extractable GTP is converted into GDP (i.e., 3 mM/min; ref. 8) and a k_{cat} of 50-250 min^{-1}, the amount of transducin that is (at least) present in intact ROSs can be estimated to be in the range of 0.6 to 3 mM, i.e., comparable to the known rhodopsin concentration of about 3 mM.[8,20] (N.B. For this estimation, we neglected the influence of the GMP concentration being about 60 μM).[20]

Implicitly, we assume that only a G_α* subunit to which a triphosphate is bound is able to activate the effector system (E becomes E*, cf. Fig. 5.1). This has been inferred from biochemical and mutation data (e.g., refs. 15,19-21 and refs. *loc. cit.*). Hence, GTP hydrolysis results in an uncoupling from the effector system and the disappearance of the G_α*-mediated E* state. Effector coupling occurs both at switch and non-switch regions of G_α* and probably also at the membrane-facing side of G_α*, since some effector systems are integral membrane proteins or membrane attached. Also G_α*·p[NH]ppG is known to activate E (e.g., ref. 20). The presence of a triphosphate group induces a closed nucleotide binding site,[15,22] which might actually be a structural prerequisite for effector binding. Since effector molecules are considered to stabilize the G_α*·GTP state (see previous paragraph), the presence of effectors could also stimulate the dissociated state of the G_α subunit and thereby also the classical GTP/GDP exchange mechanism. Note that this results exactly in an overall increase in GTP hydrolysis. However, once the trimer is regenerated, the exchange mechanism stops and effector coupling becomes barred. The role of the receptor is thus to aid in the GTP synthesis reaction with subsequent dissociation of the trimer. Once the G_α is dissociated, the amount of effector molecules and the (local) GTP concentration designates amplification level 3 activity.

GTP Hydrolysis and Inactivation

GTPase activity of G_α*·GTP results in the formation of G_α with a possible release of GDP and P_i. Subsequently, it can either bind another GTP molecule from the cytosol (amplification level 3 for which the presence of effector molecules and a high enough local concentration of GTP are important) leading to G_α* with effector coupling and hydrolysis activity again, or G_α can bind GDP associated with $G_{\alpha\beta\gamma}$ trimer formation, leading to inactivation. Not the hydrolysis but the recombination to the trimer represents the intrinsic deactivation of G proteins. This is in perfect accordance with the observation that $G_{\beta\gamma}$ suppresses the 'noise' generated by ligand-unoccupied receptors,[23] i.e., $G_{\beta\gamma}$ is known to lower basal activity. (N.B. Next to this intrinsic deactivation process, in ROSs an inhibitory factor for the effector is available that is independent of GTP hydrolysis, ref. 24; cf. chapter 6).

The same process of cytosolic GTP binding to dissociated G_α explains why GTP accelerates exchange rates.[16] As in the classical model, we thus assume that there exist two states of the dissociated G_α subunit: one with a closed nucleotide binding site, $G_\alpha{}^*$, and one with a more open site, G_α. The only difference between level 3 activity of the dissociated G_α subunit with the classical model (cf. Fig. 5.1) is the possibility that a dissociated G_α participates in the classical GTP/GDP exchange mechanism without trimer interference; effector molecules can speed up this exchange. In other words, the trimeric G protein is unable to be involved in GTP/GDP exchange reactions whereas a once dissociated G_α subunit is considered to have a feasible barrier of activation for this exchange reaction in the presence of effectors. If enough GTP is present in the direct surroundings of the dissociated G_α, it will cycle between the G_α and $G_\alpha{}^*$ states (level 3, Fig. 7.1). Now it can also be explained why the G_α:$G_{\beta\gamma}$ ratio can influence the overall GTP hydrolysis directly, which has clearly been demonstrated by Fung.[25] An hydrolysis optimum is found at a G_α:$G_{\beta\gamma}$ ratio of 10.[25] In the classical scheme (Fig. 5.1), the influence of this ratio is not obvious, because here deactivation is explained by one sole GTP hydrolysis on the $G_\alpha{}^*$ subunit. Because the amount of trimers designates GTP binding in the classical model, this implies that the more trimers are present, the higher the GTP binding and, in the end, the higher the GTP hydrolysis. The classical model thus expects that a G_α:$G_{\beta\gamma}$ ratio yielding as many G_α subunits in the trimeric $G_{\alpha\beta\gamma}$ form as possible (i.e., ratio of 1 or even less) will lead to a, as high as possible, overall GTP hydrolysis. In our scheme, however, the deactivation process consists of the recombination to the trimeric state, leading to a clear role for the $G_{\beta\gamma}$ in the overall GTP hydrolysis by the G protein. The available concentration of $G_{\beta\gamma}$ affects the time during which G_α is dissociated.

Since a dissociated G_α is suggested to be involved in a GTP/GDP exchange mechanism (amplification level 3, Fig. 7.1) and therefore binds and hydrolyzes cytosolic GTP (provided a high enough local concentration), the amount of $G_{\beta\gamma}$ directly influences overall hydrolysis by dissociated G_α subunits.

Binding of *cytosolic* GTP to dissociated G_α also explains why Robinson and Hagins[8] observe during the first 4 s after activation of rhodopsin a small increase in labelled GDP which is matched by an equal loss in pre-incubated labelled GTP. Since hydrolysis is accompanied by a slow nucleotide release with a half time of 10 s,[8] synthesis will be favored over hydrolysis in this short 4 s time period. Labelled GDP can be released only after a GTP synthesis reaction from cold pre-bound GDP has occurred, followed by a slow hydrolysis reaction. Only then can labelled GTP bind and subsequently be hydrolyzed, releasing labelled GDP. (N.B. Since triphosphate guanosines are bound tightly to G_α^*, an exchange of GTP for [^{3}H]GTP seems unlikely). Therefore, within the first 4 s after activation only small amounts of labelled GDP are expected to be released.

In the experiments reported by Robinson and Hagins,[8] GTP hydrolysis starts to catch up with synthesis after 4 s. In the period between 4 and 64 s, the rise in GDP levels even equals the fall in GTP levels. This is probably caused by a depletion of the $G_{\alpha\beta\gamma}$ trimer pool, since Chabre et al[5] have indicated that the in vivo kinetics for trimer regeneration is very slow compared to the activation and deactivation processes. Trimer association is shown to depend on G_γ prenylation, and lipids attached to G_α and $G_{\beta\gamma}$ may also be one of the determinants in the (relatively slow) trimer formation process.[26,27] In the absence of the trimeric form, the GTP synthesis reaction does not occur. Our model (Fig. 7.1) is thus adapted with an amplification at level 3 to account for the observed guanosine levels depicted in Figure 6.1, and to explain the influence of the G_α:$G_{\beta\gamma}$ ratio on basal activity and hydrolysis, and also for the known GDP/GTP exchange patterns (for exchange patterns in relation to non-hydrolyzable GTP analogues, *vide infra*); dissociated G_α apparently does not readily recombine to $G_{\alpha\beta\gamma}$ and can thus be activated to G_α^* again by a sufficient concentration of cytosolic GTP, aided by the presence of effector molecules.

It is noted that in our model (Fig. 7.1) G proteins are activated more than once. Not only is there the classical possibility of reactivation following the first round, i.e., after trimer regeneration the G protein can again be activated by R*, but also one $G_\alpha{}^*$ can be reactivated by cytosolic GTP aided by the presence of effector molecules. We have indicated above that it is likely that only a $G_\alpha{}^*$ with a triphosphate group attached to it, and thus with a closed nucleotide cleft, can bind to and activate an effector molecule (E becomes E*). But effector molecules are also known to be regulated by $G_{\beta\gamma}$ subunits. This means that effector regulation depends not only on the lifetime of the $G_\alpha{}^*$·GTP complex, but also on the time period that the G protein has actually been dissociated. After releasing the hydrolysis products GDP and P_i, another GTP can be bound (half-time of 10s)[8], which in turn yields the closed nucleotide cleft of $G_\alpha{}^*$. Only when a new GTP molecule from the cytosol is bound does ($G_\alpha{}^*$) activity return. In our model, the deactivation of the G protein lasts therefore longer than the subsecond scale indicated by Chabre et al.[5] However, the same authors[5] do not consider the possibility that effectors can also be regulated by $G_{\beta\gamma}$ dimers (which are available for such regulation until they are recombined with a G_α subunit to the trimer), whereas they do indicate that trimer regeneration takes many seconds. The G protein's intrinsic deactivation process thus occurs at the trimer regeneration time scale, and therefore we question the alleged subsecond deactivation of the G protein and emphasize that only trimer regeneration prevents both $G_\alpha{}^*$ and $G_{\beta\gamma}$ from interacting with the effector system. Next to the G protein's intrinsic deactivation process, a GTP hydrolysis-independent inhibitory factor has been indicated by Erickson et al.[24] (*vide supra*). We claim that the $G_{\beta\gamma}$ interaction with the effector could actually be the origin of this GTP hydrolysis-independent inhibitory factor. In our model, the physiological need for a subsecond deactivation of the effector system in ROSs is accounted for by this alleged $G_{\beta\gamma}$ inhibition.

The recombination of G_α with $G_{\beta\gamma}$ to form $G_{\alpha\beta\gamma}$ lasts many seconds.[5] This phenomenon might actually be accompanied by the phosphorylation of the G_β's histidine, necessary for the GTP synthesis; G_β phosphorylation might also play a role in effector regulation by $G_{\beta\gamma}$ dimers. Kowluru and co-workers indicate that the trimeric form of G is required for G_β phosphorylation.[4] If there are less $G_{\beta\gamma}$ dimers than G_α subunits present, GTP binding from the cytosol with subsequent hydrolysis is favored over trimer regeneration (an hydrolysis optimum is found at a G_α:$G_{\beta\gamma}$ ratio of 10).[25] The actual occurrence of dissociated G_α coupling to either GTP or $G_{\beta\gamma}$ also depends on the GTP/GDP ratio in the immediate vicinity of the G protein. High ratios favor GTP binding and accelerate GTP hydrolysis. From Le Chatelier's principle it follows that $G_{\alpha\beta\gamma}$ trimer formation in its turn can be stabilized by increasing GDP levels (decreasing GTP/GDP ratio), since both G_α (the more G_α the higher the chance of G_α with $G_{\beta\gamma}$ coupling) and trimers display high affinity for the diphosphate. Thus, changes in the GTP/GDP ratio in the vicinity of the G protein are expected to influence the preference for either the GTP hydrolysis or the inactivation process. Large changes in the overall GTP/GDP ratio will only be observed when the G protein is abundant, as in rod outer segments. In most other cells, however, only very small concentrations of G protein will be present and changes in overall guanosine levels are expected to be much less pronounced. Local changes in GTP/GDP ratios in the surroundings of the G protein can, however, be profound, although very difficult to measure experimentally.

Within our model, the Gibbs energy released in the exergonic GTP hydrolysis reaction is used to reduce the affinity of G_α^* for the hydrolysis products GDP and P_i and to change G_α^* into G_α. The release of these hydrolysis products is made possible through opening of the nucleotide binding cleft (see postulates, chapter 8). Analogies with the ATP synthases become again apparent, since these enzymes also use Gibbs energy to induce conformational changes and to release bound nucleotide.

7.2 EXCHANGE PATTERNS AND NON-HYDROLYZABLE GTP ANALOGUES

NON-HYDROLYZABLE GTP ANALOGUES

The implications of our model for the effects of non-hydrolyzable GTP analogues will now be discussed. Signal amplification measured as e.g., cAMP production could actually be achieved by only one dissociated G_α subunit when high levels of GTP, low amounts of $G_{\beta\gamma}$, and effector molecules are present to stabilize the active $G_\alpha{}^*$ form. Under these conditions, coupling to $G_{\beta\gamma}$ will occur rarely and effector coupling will be enabled. After GTP hydrolysis and release of the hydrolysis products GDP and P_i, another GTP or GDP with a $G_{\beta\gamma}$ can couple, leading to the prolongation of $G_\alpha{}^*$·GTP or to the regeneration of the trimeric $G_{\alpha\beta\gamma}$·GDP form, respectively. However, in the presence of non-hydrolyzable GTP analogues (p[NH]ppG or GTPγS) a constantly activated state of the G protein is obtained. Since neither p[NH]ppG nor GTPγS can be hydrolyzed by $G_\alpha{}^*$, no Gibbs energy is available for either the transformation from $G_\alpha{}^*$ into G_α, or nucleotide release, and trimer regeneration is prohibited. In addition, these non-hydrolyzable guanosine analogues are able to decouple G_α subunits from the G protein in test-tubes *in the absence of any receptor* so that an increase in basal activity is already observed when these compounds are introduced at nM concentrations[4] and the Mg^{2+} concentration is sufficiently high. How do we explain the parallel exchange rates between GDP and p[NH]ppG (or GTPγS) in light of our new model (Fig. 7.1)? In our model, p[NH]ppG can act at three levels, GTPγS at four:

(i) p[NH]ppG can act on amplification level 3, where the number of released GDP molecules parallels the amount of newly bound GTP from the cytosol. If p[NH]ppG is present, (pre-bound) GDP-P_i is first synthesized to GTP, $G_\alpha{}^*$·GTP dissociates and after activating one or more effector molecules, $G_\alpha{}^*$ releases GDP and P_i and is transferred into G_α, which in turn binds p[NH]ppG to become $G_\alpha{}^*$·p[NH]ppG.

(ii) Fung and Stryer[16] assume that p[NH]ppG can exchange for GTP bound to $G_\alpha{}^*$. In our model, this results in two ways for intervention in the activation cycle (Fig. 7.1): (a) GTP is expelled from $G_\alpha{}^*$ directly after its formation from GDP and P_i at the trimer, making subsequent hydrolysis to GDP and P_i not a prerequisite for amplification, or (b) GTP is expelled from $G_\alpha{}^*$ which already acts on level 3, i.e., when the dissociated $G_\alpha{}^*$ has bound a cytosolic GTP. In this way p[NH]ppG binding no longer exactly parallels (pre-bound) GDP release. Options (a) and (b) probably do not proceed to a proper extent, because the closed nucleotide binding site of G_α* does not easily allow exchange of the bound nucleotide.

(iii) The "classical exchange mechanism": p[NH]ppG exchanges for pre-bound GDP from the $G_{\alpha\beta\gamma}$ trimer and causes G_α*·p[NH]ppG formation without any receptor intervention if Mg^{2+} concentrations are high enough. However, in our model this first requires the dissociation of the G_α·GDP·Mg^{2+} complex from the trimer. We consider the role of Mg^{2+} in weakening G_α-$G_{\beta\gamma}$ interactions. In this way, p[NH]ppG is able to activate G proteins directly, up to a considerable level. Here, too, p[NH]ppG binding exactly parallels GDP release. Samama et al[9] assume that in the presence of p[NH]ppG all G protein is dissociated, and this 'all dissociated' state is partially accomplished without receptor intervention in case Mg^{2+} concentrations are high enough.

(iv) For GTPγS there is an additional way (besides i-iii) to intervene within our model. As mentioned earlier GTPγS is able to thiophosphorylate the G_β subunit of the G protein that couples to the GPCR fMet-Leu-Phe. The thiophosphate group can then be transferred to GDP, again resulting in GTPγS.[1] Thus, GTPγS now interferes with our scheme on level 1 (Fig. 7.1), i.e., the GTP synthesis reaction; *in casu* a GTPγS synthesis reaction. GTPγS activates the G protein pool up to a two-fold higher level than p[NH]ppG.[1] This additional possibility (iv) for GTPγS to activate the G protein might explain this two-fold enhancement. In this fourth way (iv), which has been observed experimentally, the classical exchange pattern will not occur.

GDP/GTP Exchange Patterns

In the previous subsection the observed exchange patterns for non-hydrolyzable GTP analogues are explained within our new model. In this subsection a similar objective is aimed at for GTP/GDP exchange patterns.

The overall scheme (Fig. 7.1) starts with GTP synthesis from pre-bound GDP yielding $G_{\alpha\beta\gamma}$·GTP (level 1). This process proceeds on a millisecond time scale. After G_{α}*·GTP dissociation, GTP is hydrolyzed and the products are released (with $t_{1/2}$ of 10 s; ref. 8). Cytosolic GTP can then bind to the dissociated G_{α} with an accompanied release of GDP (level 3). This is thus the same exchange mechanism which has been assumed to occur on trimeric G protein activation in the classical model (Fig. 5.1). Although we question the GTP/GDP exchange within the trimer in which the nucleotide binding site is screened from the environment, we do, however, consider this exchange mechanism to be feasible for a dissociated G_{α} in the presence of effectors. In addition, we added amplification possibilities for one dissociated G_{α} to participate in more than one exchange step. Thus, all pre-bound GDP (to the trimeric G protein) will first be converted into GTP and after hydrolysis released from G_{α} as GDP and P_i. Only then can cytosolic GTP bind. This process will result overall in parallel exchange patterns of pre-bound GDP for cytosolic GTP. The GTP synthesis with subsequent hydrolysis and release of the hydrolysis products should occur to some extent within the first seconds after activation in order to agree with the data concerning minor labelled GDP release stemming from tritiated pre-incubated GTP within 4 s after bleaching ROSs as observed by Robinson and Hagins[8] (*vide supra*).

Further support for amplification on level 3 in Figure 7.1 can be obtained from the observation that $G_{s\alpha}$ subunits, in which the wild-type Ala[366] is replaced by a Ser[366], display a higher rate of GTP hydrolysis (up to 20 times, Liri et al;[28] this mutation is reported in patients who suffer from testotoxicosis, a form of precocious puberty). A366S mutation also results in thermolability of the $G_{s\alpha}$, causing pseudohypoparathyroidism type Ia (cf. ref. 28 and refs. therein for further details). The higher rate of hydrolysis in these A366S mutants has been ascribed to a more rapid release of GDP. This is in complete agreement with our model, since the release of the hydrolysis products is considered to be the rate limiting step in the exchange mechanism for the dissociated G_{α}. If the GDP release is facilitated, subsequent cytosolic GTP binding is facilitated and higher effector stimulation as well as enhanced

GTP hydrolysis activity is to be expected, compared to the wild-type, which is in total agreement with the experimental results reported by Liri et al.[28]

Amplification level 3, i.e., the GTP/GDP exchange reaction at the dissociated G_α subunit, is essential to explain the reported concentrations of guanine nucleotide levels for ROSs monitored after the activation by light and the "classical" GTP/GDP exchange patterns. But purified G_α subunits are labile in the absence of nucleotides (Misha Freissmuth, personal communication; and ref. 28 as well as refs. *loc. cit.*). This raises the question of whether or not amplification level 3 can occur at all, since during this exchange process empty dissociated G_α subunits are present (N.B. A persistently empty nucleotide binding site at $G_{s\alpha}$ is claimed to be the origin of the thermolability of the aforementioned A366S mutants[28]). In this respect it is worth while to point out that we have claimed amplification level 3 to exist in an intact cell situation. We have already indicated the importance of effector molecules for the GTP/GDP exchange reaction at G_α, and pointed to the fact that effector molecules are known to be able to speed up the GTP hydrolysis rate of the G protein (refs. 17-19 and refs. *loc. cit.*). In this respect a comparison with negative regulators of Ras proteins can be made. These negative regulators are also known as GTPase activating proteins (GAPs).[29-35] GAPs preferentially interact with the active Ras·GTP complex and are known to accelerate the otherwise slow intrinsic GTP hydrolysis of Ras by a factor of 10^5. We have outlined that the hydrolysis of GTP at $G_\alpha{}^*$ is necessary for a possible subsequent release of the hydrolysis products to enable the coupling of another GTP molecule thereby prolonging the active $G_\alpha{}^*$ state. Here, another similarity with the Ras system is noticed, i.e., positive regulators of Ras proteins are known, called guanine nucleotide exchange factors (GEFs). GEFs promote the release of GDP from Ras eventually resulting in the formation of the active Ras·GTP complex.[30,35-37] The mechanism is as follows: GEFs associate with the inactive Ras·GDP complex. After GDP dissociation the GEF binds to an empty Ras protein, thereby stabilizing the empty state. An empty Ras in turn enables GTP binding, leaving Ras in its active GTP-bound form. The GEF, having low affinity for the Ras·GTP complex, dissociates. (Because of the discrepancies between in vitro and in vivo data, other cellular factors could influence the affinity of GEFs for Ras·GDP complexes.) In the case of G proteins, it is likely that effector molecules (or other cellular factors) act as GAPs or GEFs, since in our opinion both the phenomena of GTP hydolysis and GDP dissociation will yield amplification at level 3 in scheme 7.1. If GTP hydrolysis delivers enough energy to release the hydrolysis products, the next GTP molecule can bind to G_α, and the active $G_\alpha{}^*$ state continues to exist. Hence, the faster GTP hydrolysis is, the faster the cycling at level 3 will be. Furthermore, if the relatively slow release of hydrolysis products is accelerated by the presence of a cellular GEF, again cytosolic GTP is enabled to bind to the dissociated G_α, again resulting in faster cycling at level 3. Therefore, the fact that purified empty G_α subunits are labile does not interfere with our amplification level 3, although it must be stressed that this level is expected in intact cells only, with the presence of both GAP- and GEF-like proteins. Also activated G protein-coupled receptors (R^*) are known to accelerate the GTP/GDP exchange at purified G_α subunits (in the *absence* of $G_{\beta\gamma}$). R^* can thus be considered as a GEF for the dissociated G_α subunits and in intact cells the presence of R^* contributes to the amplification at level 3 in Figure 7.1.

7.3 INTERPRETATION OF THERMODYNAMICAL DATA

The classical model for G protein activation (GDP exchange for GTP) interprets previously measured thermodynamical data in a different way than our model (GTP synthesis) implies. Stryer[20] (classic model) states that the vision excitation cycle for rhodopsin is driven by GTP hydrolysis and not by the energy of the absorbed photon. However, the exergonic $G_\alpha{}^*$ dissociation reaction (Eq. 6.1, K_{eq} 100 M; *vide supra*) is not considered. Furthermore, for non-hydrolyzable GTP analogues large amplification is also observed, pointing to an exergonic reaction, although there is no energy from an hydrolysis reaction available.

As mentioned earlier, experimental data on thermodynamics of receptor activation and the associated half time of GTP in rod outer segments[5-8] can only be brought in accordance when it is assumed that $G_\alpha{}^*$ dissociation occurs on a millisecond time scale accompanied by heat release, while GTP hydrolysis is a much slower process ($t_{1/2}$ ~ 10s). Experimental verification seems to be possible, since for both GTP and its non-hydrolyzable analogues a similar heat release should be observed in our model, whereas if the heat release is purely associated with GTP hydrolysis, such a release should be absent when only non-hydrolyzable analogues are used (see also chapter 6).

7.4 DESENSITIZATION

Finally, we briefly address the phenomenon of desensitization within the classic model and our new model. Not only the heterotrimeric protein $G_{\alpha\beta\gamma}$·GDP·P_i has affinity for the cytoplasmic loops of the GPCR, but kinases are also reported to couple to and subsequently phosphorylate cytoplasmic residues of the agonist-occupied receptor (for a review see e.g., ref. 38, and refs. *loc. cit.*). This intracellular loop phosphorylation, occurring within seconds to minutes, results in a decreased efficiency of receptor interaction with the G protein, and is called desensitization. Since both kinases and G proteins bind to the agonist-occupied receptor, there might be a competition for binding to the activated receptor R*. Kinase binding might stabilize the R* state, just as G proteins do, which could lead to enhanced agonistic high affinity binding. Also $G_{\beta\gamma}$ subunits are reported to aid in kinase binding to agonist-occupied receptors (e.g., ref. 39). Within the classic model and our new model, this implies that enhanced $G_{\beta\gamma}$ levels increase GPCR phosphorylation. In our model, it is tempting to suggest a possible role for the phosphorylated G_β's histidine in this process.

Since phosphorylated G_β subunits can synthesize free GTP from free GDP,[4] another potential pathway exists in our new model that could lead to a decreased efficiency of G protein activation by GPCRs. Since free GDP can dephosphorylate G_β, in the case that a phosphorylated G_β is truly necessary for G protein activation, the G protein becomes inactivated when high concentrations of free GDP are available to phosphorylated G_β. Therefore, the [GTP]/[GDP] ratio has a putative influence on G protein activation by regulating the phosphorylation state of G_β. If the GTP pool becomes exhausted by G protein activation, the resulting local [GTP]/[GDP] ratio could prohibit further activation. Only when this ratio reaches a certain threshold value does G protein activation return. Re-establishing this local ratio to the situation of a cell in its resting state may take some time, which consequently leads to desensitization.

7.5 SUMMARY

A new mechanism for G protein activation is presented in Figure 7.1. The major adjustments to the "classical" scheme (depicted earlier in Fig. 5.1) are (i) a GTP synthesis step at the trimer yielding G_α*·GTP and $G_{\beta\gamma}$ and (ii) the possibility of amplification at the GTP/GDP exchange mechanism for a dissociated G_α subunit in an intact cell situation. Within this scheme, we claim to offer an explanation for the experimental findings reported in relation to G protein action such as GTP/GDP exchange rates, direct influences of the G_α:$G_{\beta\gamma}$ ratio on the basal activity and on the GTP hydrolysis rates, phosphorylation reactions inside G proteins, nucleotide levels measured in ROSs, and a correct thermodynamical description of the energy input necessary for amplification processes.

In the next chapter we will present this new mechanism by means of five postulates incorporating bioenergetics into G protein activation mechanisms.

References

1. Kaldenberg-Stasch S, Baden M, Fesseler B et al. Receptor-stimulated guanine-nucleotide-triphosphate binding to guanine-nucleotide-binding regulatory proteins. Eur J Biochem 1994; 221:25-33.
2. Wieland T, Kaldenberg-Stasch S, Fesseler B et al. Regulation of G protein function by phosphorylation. Can J Physiol Pharmacol 1994; 72:S5.
3. Wieland T, Nürnberg B, Ulibarri I et al. Guanine nucleotide-specific phosphate transfer by guanine nucleotide-binding regulatory protein β-subunits. J Biol Chem 1993; 268:18111-18118.
4. Kowluru A, Seavey SE, Rhodes CJ et al. A novel regulatory mechanism for trimeric GTP-binding proteins in the membrane and secretory granule fractions of human and rodent β cells. Biochem J 1996; 313:97-107.
5. Chabre M, Antonny B, Vuong TM. The transducin cycle in the phototransduction cascade. NATO ASI Series, Series H 1991; 52:207-220.
6. Vuong TM, Chabre M. Subsecond deactivation of transducin by endogenous GTP hydrolysis. Nature 1990; 346:71-74.

7. Vuong TM, Chabre M, Stryer L. Millisecond activation of transducin in the cyclic nucleotide cascade of vision. Nature 1984; 311:659-661.
8. Robinson WE, Hagins WA. GTP hydrolysis in intact rod outer segments and the transmitter cycle in visual excitation. Nature 1979; 280:398-400.
9. Samama P, Cotecchia S, Costa T et al. A mutation-induced activated state of the β_2 adrenergic receptor. Extending the ternary complex model. J Biol Chem 1993; 268:4625-4636.
10. Rodbell M. Signaltransduktion: Die Entwicklung einer Theorie (Nobel-Vortrag). Angew Chem 1995; 107:1549-1558.
11. Strader CD, Dixon RAF, Cheung AH et al. Mutations that uncouple the beta-adrenergic receptor from G_s and increase agonist affinity. J Biol Chem 1987; 262:16439-16443.
12. Lee NH, Kerlavage AR. Molecular biology of G protein-coupled receptors. Drug News & Perspectives 1993; 6:488-497.
13. Hausdorff WP, Hnatowich M, O'Dowd BF et al. A mutation of the β_2 adrenergic receptor impairs agonist activation of adenylate cyclase without affecting high affinity agonist binding. J Biol Chem 1990; 265:1388-1393.
14. Ernst OP, Hofmann KP, Sakmar TP. Characterization of rhodopsin mutants that bind transducin but fail to induce GTP nucleotide uptake. J Biol Chem 1995; 270:10580-10586.
15. Lambright DG, Noel JP, Hamm HE et al. Structural determinants for activation of the α-subunit of a heterotrimeric G protein. Nature 1994; 369:621-628.
16. Fung BK-K, Stryer L. Photolyzed rhodopsin catalyzes the exchange of GTP for bound GDP in retinal rod outer segments. Proc Natl Acad Sci USA 1980; 77:2500-2504.
17. Kleuss C, Raw AS, Lee E et al. Mechanism of GTP hydrolysis by G protein α subunits. Proc Natl Acad Sci USA 1994; 91:9829-9831.
18. Neer EJ. Heterotrimeric G proteins: organizers of transmembrane signals. Cell 1995; 80:249-257.
19. Conklin BR, Bourne HR. Structural elements of G_α subunits that interact with $G_{\beta\gamma}$, receptors and effectors. Cell 1993; 73:631-641.
20. Stryer L. Cyclic GMP cascade of vision. Ann Rev Neurosci 1986; 9:87-119.
21. Gilman AG. G proteins: Transducers of receptor-generated signals. Annu Rev Biochem 1987; 56:615-649.
22. Rens-Domiano S, Hamm HE. Structural relationships of heterotrimeric G proteins. FASEB J 1995; 9:1059-1066.
23. Birnbaumer L, Mattera R, Yatani A et al. Recent advances in the understanding of multiple roles of G proteins in coupling of receptors to ionic channels and other effectors. In: Moss J, Vaughan M,

eds. ADP-ribosylating toxins and G proteins, Insights into signal transduction. Washington DC: Am Soc Microbiol, 1990:225-266.
24. Erickson MA, Robinson P, Lisman J. Deactivation of visual transduction without guanosine triphosphate hydrolysis by G protein. Science 1992; 257:1255-1258.
25. Fung BK-K. Charaterization of transducin from bovine retinal rod outer segments. Separation and reconstitution of the subunits. J Biol Chem 1983; 258:10495-502.
26. Birnbaumer L, Birnbaumer M. Signal transduction by G proteins: 1994 edition. J. Receptor & Signal Transduction Research 1995; 15:213-252.
27. Casey PJ. Protein lipidation in cell signalling. Science 1995; 268:221-225.
28. Liri T, Herzmark P, Nakamoto JM et al. Rapid GDP release from $G_{s\alpha}$ in patients with gain and loss of endocrine function. Nature 1994; 371:164-168.
29. Bernards A. Neurofibromatosis type 1 and Ras-mediated signalling: filling in the GAPs. Biochim Biophys Acta 1995; 1242:43-59.
30. Boguski LA, McCormick F. Proteins regulating Ras and its relatives. Nature 1993; 366:643-654.
31. Marshall MS. The effector interactions of $p21^{ras}$. Trends Biochem Sci 1993; 18:250-254.
32. Marshall MS. Ras target proteins in eukaryotic cells. FASEB J 1995; 9:1311-1318.
33. McCormick F. Ras GTPase activating protein: signal transmitter and signal terminator. Cell 1989; 56:5-8.
34. Pronk GJ, Bos JL. The role of $p21^{ras}$ in receptor kinase signalling. Biochim Biophys Acta 1994; 1198:131-147.
35. Wiesmüller L, Wittinghofer F. Signal transduction pathways involving Ras. Cellular Signalling 1994; 6:247-267.
36. Feig LA. Guanine nucleotide exchange factors: a family of positive regulators of Ras and related GTPases. Curr Opinion Cell Biol 1994; 6:204-211.
37. Quilliam LA, Khosravi-Far R, Huff SY et al. Guanine nucleotide exchange factors: activators of the Ras superfamily of proteins. BioEssays 1995; 17:395-404.
38. Collins S. Molecular structure of G protein-coupled receptors and regulation of their expression. Drug News & Perspectives 1993; 6:480-487.
39. Mochly-Rosen, D. Localization of protein kinases by anchoring proteins: a theme in signal transduction. Science 1995; 268:247-250.

CHAPTER 8

Principles of a New Molecular Mechanism for Signal Transduction

If our model has to stand the test of critics, it should of course explain all findings in connection with signal transduction of GPCRs and provide more insights in the several processes involved than the classical theory does. In the previous chapter, we have described schematically a new model for G protein activation. We will end by presenting five postulates on the manner in which GPCRs and G proteins act in this new model, incorporating the knowledge and new insights described in the previous chapters. In chapter 9, we will indicate avenues for further work.

8.1 POSTULATE 1: GPCRs possess a proton wire (HBC) to translocate protons from the extracellular space across the membrane.

GTP synthesis is a crucial step in the catalytic activation cycle as presented in Figure 7.1. To enable GTP synthesis, the presence of a protonmotive force generated by a primary pump is important. We therefore postulate the presence of a hydrogen-bonded chain (HBC) acting as a proton wire within the activated receptor protein, which translocates the protons provided by (a) primary pump(s) across the membrane (cf. postulate 5).

Constitution of the Proton Wire

In the resting state, the receptor's proton wire is blocked either by a hydrophobic gate or by the lack of one entity in the wire, resulting in a mainly inactive chain (only basal activity is present) and preventing the protein from being (constitutively) active. In general, the function of an agonist is to open the hydrophobic gate or to bind to that position where an entity is missing in the proton wire, thereby restoring its proton translocation ability (Fig. 8.1). For the G protein-coupled opsins, the 'ligand' is light (Fig. 8.2).

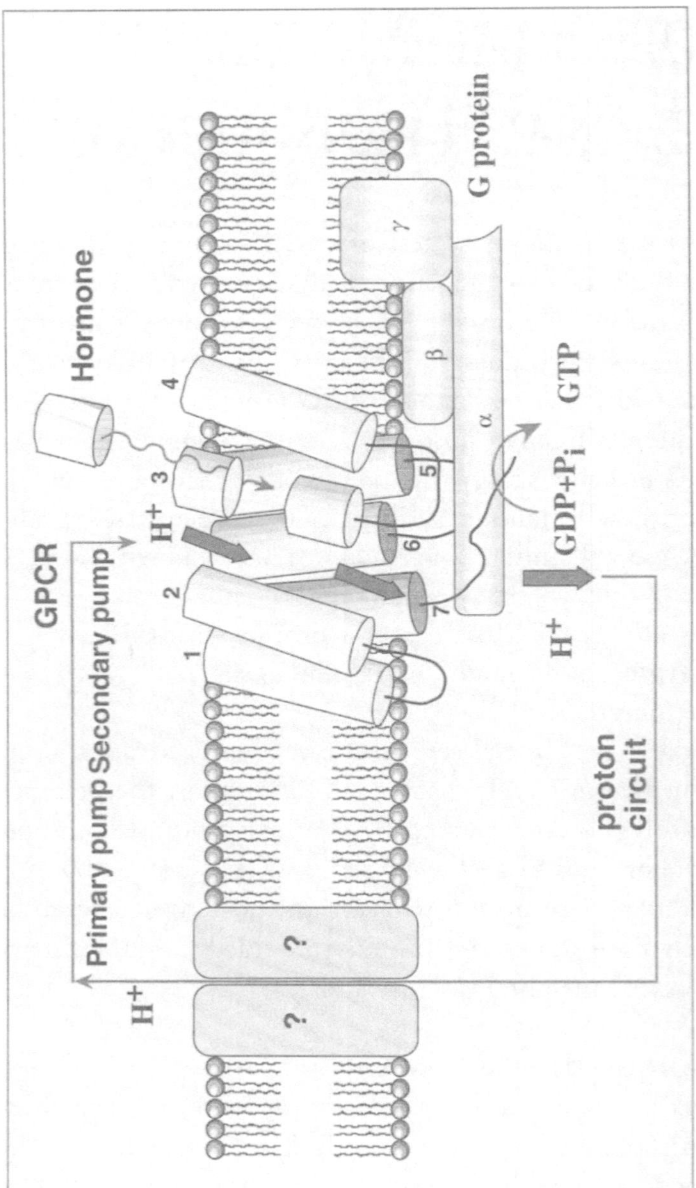

*Fig. 8.1. Binding of an agonist restores the HBC, resulting in a ternary complex (HR*G). Topiol's deletion model[1] is followed. The proton gradient, generated by an as yet unknown primary pump, can be used to translocate protons via the GPCR and to synthesize GTP from GDP and P_i at the catalytic G_α subunit of the G protein mediated by phosphorylation and dephosphorylation of a histidine residue located on the G_β subunit.[2,3] (Reprinted with permission from: Nederkoorn PHJ, Timmerman H, Donné-Op den Kelder GM. Does the ternary complex act as a secondary proton pump and a GTP synthase? Trends Pharmacol Sci 1995; 16:156-161.)*

After ligand-induced receptor activation (Fig. 8.1), the proton wire exists in e.g., an OH--OH--OH state (OHs represent proton translocating amino acids or water molecules), while after the proton translocation (the hop) the chain occurs in a HO--HO--HO state (Fig. 1.2). For this process to occur the indicated polar character of the receptor cavity[5] is an absolute necessity. Sealfon et al[6]

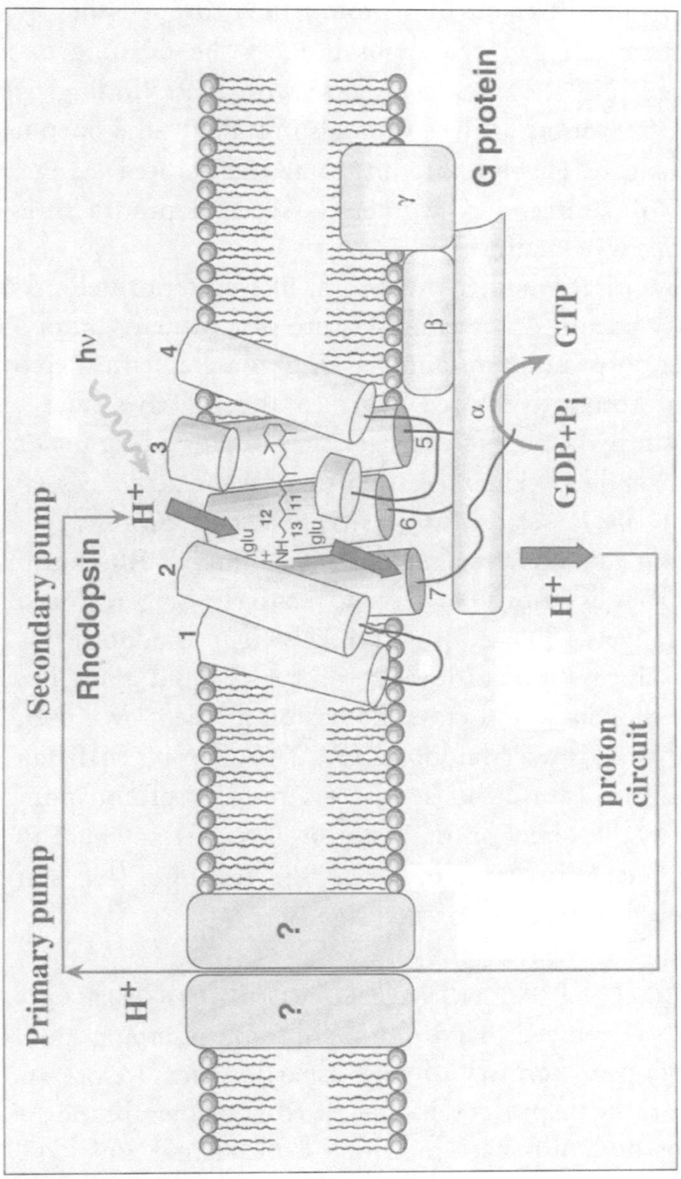

Fig. 8.2. Upon light activation, G protein-coupled rhodopsin acts as a secondary pump. Thereby GTP is synthesized from pre-bound GDP and P_i at the trimeric G protein. Here, the prolonged opening of a hydrophobic gate enables signal amplification. Agonists can thus activate GPCRs in two ways: (i) via direct interaction with the HBC (cf. Topiol's deletion theory in Fig. 4.2 and 8.1) or (ii) via opening of a hydrophobic gate. The primary pump activity is measured with the Cytosensor Microphysiometer[70,71] (see text for further details). (Reprinted with permission from: Nederkoorn PHJ, Timmerman H, Donné-Op den Kelder GM. Does the ternary complex act as a secondary proton pump and a GTP synthase? Trends Pharmacol Sci 1995; 16:156-161.)

have already suggested the presence of a common hydrogen-bonding network underlying GPCR activation by an agonist.

Activating the GPCR Rhodopsin (Fig. 8.2)

Rhodopsin is a G protein-coupled receptor with a large functional homology to bacteriorhodopsin. Both are light-activated transmembrane proteins and have retinal as their chromophore. For bacteriorhodopsin it is known that light-induced conformational changes in retinal result in a proton flow through the protein. Dencher et al[7] suggest that light induces the opening of a hydrophobic gate (Fig. 2.1). In accordance with these findings we suggest that the activation of rhodopsin also results in the opening of a hydrophobic gate, enabling protons to be translocated. As early as the 1980s Hofmann and co-workers[8] associated proton transfers with rhodopsin activation.

The question arises whether rhodopsin, like bacteriorhodopsin, acts as a primary pump. An essential feature of a primary pump is that after activation protons are pumped *against* a gradient, thereby generating a protonmotive force (Δp). In the inactive state of bacteriorhodopsin hydrophobic gates are closed, preventing uncontrolled proton backflow.[7] The relaxation time of the activated state (*13-cis* state) of BR is very short (ms), preventing protons from flowing back immediately after having been pumped. Rhodopsin, on the other hand, is known to have a relatively long relaxation time (minute range[9]). Hence, it is unlikely that rhodopsin is a primary pump since the translocated protons would then have enough time to flow back and erase the proton gradient. We therefore conclude that activated rhodopsin (R*, i.e., *all-trans* MII-state, cf. ref. 10) bound to transducin acts as a secondary proton pump (protons move in the direction of a proton gradient) and that the relatively long relaxation time of R* to the resting *11-cis* state enables a huge amplification of the signal.

A Minimal Proton Pumping Model for Rhodopsin

In chapter 2 we showed that a minimal proton pumping model for bacteriorhodopsin consists of two aspartic acids (Asp^{85} and Asp^{96}), which are both present in the third transmembrane helix (TM3), and the Schiff base in which both retinal and Lys^{216}

participate. Bovine rhodopsin possesses—besides a similar Schiff base between retinal and a lysine residue—two glutamate residues (Glu^{113}, Glu^{122}) present at relatively the same positions with respect to the chromophore retinal as Asp^{85} and Asp^{96} in BR.[11] Mutation studies on rhodopsin (Glu to Gln mutations) leading to a large reduction in the activated (R*) 380nm species,[12] provide evidence that Glu^{113} and Glu^{122} could indeed have a similar function in proton conduction as the two aspartates in BR.

Additional evidence for a related mechanism in these two proteins can be deduced from resonance Raman studies reported by Deng et al,[13] who suggest that in both proteins one or more water molecules are present near the Schiff base. These waters of hydration could directly be involved in a proton transfer process. Also, from experiments conducted by Arnis and Hofmann[14] it is known that photorelease of the Schiff base proton in rhodopsin is followed by proton uptake from the aqueous phase. Moreover, the release of G protein-dependent H^+ has actually been observed.[15] Even more evidence can be inferred from the phototaxis receptor sensory rhodopsin SR-I, which in the absence of signal transducers acts as a primary pump, but reverses pumping in the presence of transducer molecules.[16] Probably, the transducer molecules absorb the proton(s) released from the Schiff base, thereby preventing primary proton pump activity of SR-I and converting the system into a secondary pump. The analogy with rhodopsin now becomes evident: on activation, proton(s) are introduced into the G protein. Sakmar and co-workers[12] have shown for rhodopsin that both Glu^{134} and Arg^{135} play a role in G protein activation and that these two amino acids can therefore be considered as a part of the proton wire.

Synthetic Peptides and the HBC

The GDP-bound form of the G protein has a high affinity for the activated states R* and HR*. G_α^*·GTP has a low affinity for R* and HR* but a high affinity for the effector system. Synthetic peptides, mimicking short portions of the intracellular loops of GPCRs (refs. 17-21), cause G_α^* dissociation and stabilize G_α^*·GTP independent of the signalling state of the receptor. Therefore, it has been suggested that these peptides compete with R* for binding to $G_{\alpha\beta\gamma}$·GDP. Ross[22] emphasizes the importance of cationic domains in these synthetic peptides, and Birnbaumer and Birnbaumer[23] also indicate the presence of basic residues within these peptides. We suggest that these peptides can inject a proton(s) into the G protein, thereby inducing an ionic or bonding fault inside the G protein (cf. Fig. 1.2), which in turn causes G protein activation.

Possibilities for Experimental Verification of Proton Translocation

Keefer et al[24] conducted experiments to demonstrate continuous proton movements in response to agonists. These attempts however failed. The experiments were done in vesicles with purified α_{2a} adrenergic receptors and occasionally in the presence of G protein. We have two comments on the experimental set-up chosen:

(i) Ion transport across membranes can only be measured when *both* a potential *and* a circuit is present. Even when a membrane potential is present driving proton translocation, the transfer will stop if it is not compensated by an active or passive transport in the opposite (ions with similar charge) or similar (ions with opposite charge) direction, because the membrane will become polarized, preventing further ion transport; H^+ efflux can be matched by K^+ influx in the presence of valinomycin. In the proton movement experiments reported by Keefer et al[24] such a circuit is absent.

(ii) Our second comment concerns the observation of certain agonist-dependent activity in the absence of a proton gradient. We claim that this activity may be caused in a similar way to the one observed for synthetic peptides, mimicking parts of the third intracellular loop (*vide infra*).

Synthetic peptides with cationic domains mimicking parts of the intracellular domains of GPCRs are known to activate G proteins up to a significant, albeit much lower, level when compared to the activity observed after stimulation of GPCRs in intact cells (e.g., refs. 23, 25 and refs. *loc. cit.*). Within our new model, these cationic domains can be seen as potential H^+ donors, which are able to activate the G protein through the injection of protons. Hence, a Δp is not even an absolute prerequisite for G protein activation. The presence of cationic domains in the intracellular loop(s) of the GPCR is already sufficient to achieve a certain low level of G protein activation. This further explains that in cells with broken membranes, resulting in the dissipation of ionic gradients, some GPCR-induced G protein activity is still observed. In contrast, permeabilization of membranes gives rise to enhanced pumping activities of e.g., proton transporters in mitochondria (e.g., ref. 26). This results in an immediate increase in G protein activation, as observed after addition of K^+-selective toxins such as valinomycin and α-toxins from *Staphylococcus aureus*.[27,28] However, we predict that prolonged exposure to these toxins, which will eventually lead to a depletion of the energy sources available for ion pumping (Δp collapses), will ultimately result in impaired G protein activation.

Furthermore, it is recalled that the application of classical protonmotive force uncouplers, such as FCCP or CCCP,[26] results in short-circuiting the proton circuit (cf. chapter 1 and Fig. 1.1), i.e., the Δp diminishes/vanishes in the presence of these decoupler molecules.[26] The importance of the presence of a Δp for amplification of G protein activation can thus be tested experimentally by adding this kind of molecule to a pharmacological assay. First attempts in our laboratories have indeed shown that uncouplers influence effector systems (unpublished results; Rob Leurs, personal communication). When the human histamine H_2 receptor (stably expressed in Chinese hamster ovary cells) is stimulated by its endogenous ligand, the G_s protein becomes activated, resulting in an increase in cAMP production via activation of adenylate cyclase. Preliminary experiments show that in the presence of a concentration gradient of 0.1-50 μM FCCP (half-hour incubation), histamine-stimulated cAMP production significantly decreases. Since FCCP influences the ATP household of an intact cell and adenylate cyclase activity depends on the ATP concentration, direct activation of adenylate cyclase by forskolin was also monitored as a (negative) control. Pre-incubation of half an hour with FCCP displayed in some cases that the forskolin response itself was also lowered. Therefore, no definite answer can yet be given as to whether or not FCCP directly influences G protein activation. Additional experiments are required in which the time of pre-incubation with uncoupler molecules on forskolin responses is monitored. Moreover, other effector systems, less dependent on the ATP concentration inside the cell, should be considered as well.

In conclusion, we predict that the absence of a protonmotive force results in impaired G protein activity since the amplification at level 1, which is (partially) Δp driven, will decrease in the absence of a Δp (Fig. 7.1). An ideal experimental set-up must therefore closely resemble the situation as found in intact cells. We therefore suggest the use of vesicles with purified GPCRs and G proteins in combination with a controllable primary pump such as cytochrome *c* oxidase[29] or BR (ref. 30, including a detailed description of the experimental conditions) in order to verify the new model. Controllable primary pumps enable the control of the height of Δp directly (by varying either the amount of O_2 in cytochrome *c* oxidase-containing vesicles or the amount of light given to BR-containing vesicles, Δp can be regulated accordingly). Furthermore, in vesicles with high amounts of both GPCR and G protein, alterations of guanosine levels can be measured without interference from other proteins, such as guanine cyclase and cyclic nucleotide phosphodiesterase, present in ROSs.

Agonists and the Proton Wire

Agonists are postulated to induce a proton flow in the ternary complex, i.e., in HR*G. The proton wire (an HBC) runs from the P-phase (extracellular) towards the N-phase (cytosol). Agonists have a high affinity for R* in which the proton wire (but only in connection with a G protein, *vide infra*) is fully active, and a low affinity for R, respectively; the agonist therefore enhances the stability of R* relative to R (see Fig. 8.1). For a small number of GPCRs, a part of the proton wire has theoretically been elucidated.[31,32]

The agonist-induced proton flow can be established in two ways: (i) The agonist opens hydrophobic gate(s) which are subsequently filled with transient water molecules restoring a fully active proton wire. This *indirect m*echanism corresponds to the one revealed in bacteriorhodopsin and possibly also present in rhodopsin. (ii) The agonist is *directly* involved in restoring the proton wire as an essential link in the chain. Topiol's deletion model[1] is now clearly applicable: the agonist restores a proton wire (OH--gap--OH becomes OH--XH--OH; or HO--gap--HO becomes HO--HX--HO) and performs the role of the missing entity (Figs. 4.2 and 8.1). Additional evidence that ligands are indeed able to restore proton wires is obtained from a recent study on p-hydroxybenzoate hydroxylase.[33] The presence of this proton wire is dependent on substrate binding and furthermore the HBC is considered to assist the catalytic deprotonation of the 4-hydroxyl group of the substrate.[33]

Both the *indirect* and *direct* mechanism for establishing an agonist-induced proton flow can explain the apparent relationship between intrinsic agonist activity and affinity:[25] the higher the affinity of the (partial) agonist for the HBC, the better the proton wire is stabilized, and as long as the wire exists protons can be translocated, which thus relates the height of the intrinsic activity of the particular (partial) agonist to its affinity. This relationship will not hold for molecular portions which only enhance affinity but do not affect activity. Classical examples of 'high affinity molecular portions' are found within series of histamine H_2 receptor agonists; the high potency of certain agonists compared to histamine is solely due to enhanced affinity values, whereas efficacy is hardly affected (ref. 34; and refs. *loc. cit.*). The observed relation between enhanced intrinsic activity and increased affinity for the mutation-induced constitutively active β_2 adrenergic receptor[25] can thus be due to a more facile formation of the proton wire inside the mutated receptor. A direct relationship between affinity and intrinsic activity is absent when a partial agonist binds with a moderate pK_d value, but simultaneously highly improves the proton conductance of the hydrogen-bonded chain. The better the resulting proton conductance of the wire, the higher the intrinsic activity of the particular partial agonist. Binding to the wire and proton conductance of the wire are two different molecular mechanisms. Within our proton wire concept, experimentally observed phenomena, such as mutations revealing preserved G protein binding to the GPCR with retained high affinity binding of agonists yet in combination with decreased G protein activation,[35] can now be understood.

Basal Activity and the Proton Wire

Basal activity is considered to result from water molecules present in the central cavity of the receptor, which either transiently fill the agonist binding site (direct way) or open hydrophobic gate(s) (indirect way). In both ways, transient water molecules contribute to the formation of a weakly active proton wire. The higher the stabilization of R* with respect to R (an ability of constitutively active receptors, *vide supra*), the better waters of hydration are able to fill the gap or open gate(s). In other words, constitutively active receptors have a higher proton conductance and basal activity with respect to the wild type.

If we consider again Topiol's deletion model in which an agonist fills a gap in a proton wire (e.g., in OH--gap--OH), and combine this model with our ideas on basal activity, the gap in the proposed inactive state is filled with water molecules (e.g., OH--water--OH). In the active state, the waters are replaced by the agonist. The proton conductance of the agonist is higher than the (nonspecific) conductance of water; in other words, the proton conducting resistance of a hydrogen-bonded chain partly filled with transient waters is higher than the resistance of an HBC constituted with an agonist.

Antagonists, Inverse Agonists and the Proton Wire

Within the ETCM,[25] lowering of basal activity due to inverse agonists (negative antagonists) is explained by an increased stabilization of R relative to R*, resulting in a decreased occurrence of R* and a concomitant increased occurrence of R in which the third intracellular loop is shielded from the G protein. In our opinion, the ETCM does, however, not offer a satisfactory explanation for the mutation of a GPCR reported by Hausdorff and co-workers[35] in which G protein binding and high affinity binding of agonists are both preserved in combination with decreased G protein activation. The preservation of both G protein binding to the GPCR and high affinity binding of agonists point to an unaltered amount of R*. We claim that this particular mutation altered the proton conductance without affecting the R:R* ratio. In accordance with this conclusion, we propose that inverse agonists can influence G protein activation through three distinct mechanisms: (i) In the "classical" manner: inverse agonists have a higher affinity for R than for R* (e.g., ref. 36) and therefore destabilize R* relative to R. Basal activity is lowered. Since inverse agonists destabilize R*, its occurrence becomes less likely (Boltzmann distribution). Since only R* is able to couple to the G protein, less G proteins are activated. (ii) Inverse agonists can block the proton wire by introducing hydrophobic gates into the HBC which will affect the proton translocation ability negatively, resulting in a decreased basal activity. Since the introduction of a hydrophobic gate does not necessarily have to influence the R:R* ratio, G protein binding to the GPCR might not be affected. (iii) Inverse agonists could bind to R* directly and generate a highly asymmetric HBC in which the hop and turn state differ largely in energy. This creates an HBC which is unable to translocate protons (proton conductance is thus altered): even with the energy available from a Δp, either the hop state cannot be converted into the turn state or vice versa (Fig. 1.2). In this case G protein coupling to the GPCR even increases.

Neutral antagonist binding does not alter the basal activity of a GPCR.[36] The energy of the R state relative to R* is thus unchanged. In our model, the phenomenon of neutral antagonism can be explained when we assume that this class of antagonists does prevent agonist binding, but does not interfere with the proton wire in any of the three ways described above for inverse agonists. For a number of receptors the gap in the proton wire is assumed to be filled with waters (basal activity) which should thus be unaffected by neutral antagonist binding.

Mutations in GPCRs and the Proton Wire

The existence of constitutively activated mutant receptors and their possible role in diseases have been reviewed extensively (e.g., refs. 37,38, and refs. *loc. cit.*). However, for understanding the molecular mechanisms underlying G protein activation by GPCRs other mutations should be of great interest too, such as those that result in receptors that retain (or even enhance) the ability to bind agonists with high affinity but display impaired signal transduction by decreased G protein binding (e.g. ref. 39), or those that show retained high affinity binding of agonists *and* retained G protein binding but nevertheless impaired G protein activation (e.g. refs. 35; for rhodopsin an analogous mutation has been reported by Ernst et al;[40] see also chapter 5) or mutants revealing the presence of more than one agonistic binding site (e.g. refs. 41a-c,42). These mutations increase our insights into agonist binding and agonist-induced signal transduction. For example, mutation studies have revealed the presence of two different, albeit partially overlapping, agonistic binding sites at the histamine H_1 receptor.[41a-c] Also, Wiens et al[42] report different agonist binding sites at the dopamine D_2 receptor. Hence, the classical concept of one unique binding site to which all agonists bind in a similar mode and induce this one unique conformational change within the receptor is highly questionable. For both the histamine H_1 and dopamine D_2 receptors, certain mutations at the fifth transmembrane domain hardly influence agonist binding, whereas signal transduction is impaired significantly.[41a-c,42] In the particular case of the histamine H_1 receptor, residues from the fifth transmembrane α-helix positioned close to the extracellular surface were mutated.[41a-c] Within the classical concept of a conformational change inside the GPCR, which is transferred from the site of agonist binding towards the G protein at the cytoplasmic side of the membrane, this impaired signal transduction stemming from alterations in the periplasmic transmembrane domains that are not involved in agonist binding, cannot be understood.

Within the concept of an HBC, agonists can bind to (partially) different binding sites. As long as an active chain is achieved, the signal will be transmitted. Also, amino acids that are not directly involved in agonist binding can directly or indirectly (via waters of hydration) contribute to the HBC. An altered HBC can result in altered agonist binding and/or altered proton conductance. Therefore, the pharmacological behavior observed for the aforementioned mutated GPCRs can now be understood and can actually assist in elucidating the location of the HBC (or even HBCs) inside a particular GPCR.

8.2 POSTULATE 2: G_α and G_α* correspond to the E_s and E_h forms of the F_1 unit of $F_1 \cdot F_0$-ATP synthase, respectively

The two conformational forms of the G_α unit of the G protein, G_α and G_α*, are postulated to display analogies with the E_s and E_h forms of the F_1 unit of $F_1 \cdot F_0$-ATP synthase, respectively[43] (see also chapter 3). The G_α subunit, which has a rather open nucleotide binding site and high affinity for $G_{\beta\gamma}$, corresponds to the E_s form (*s* for synthesis); G_α in its E_h form (*h* for hydrolysis) is denoted G_α* which has a closed nucleotide binding site and is always separated from $G_{\beta\gamma}$.

SIMILARITIES BETWEEN THE G PROTEIN AND $F_1 \cdot F_0$-ATP SYNTHASE

Since the G_α unit in its E_s form can be associated with $G_{\beta\gamma}$, the E_s form can exist in two forms (i) dissociated with high affinity for GDP and (ii) as the heterotrimeric G protein with high affinity for GDP and P_i, just as the E_s form of ATP synthase has a high affinity for ADP and P_i. This explains why GDP and not GTP is bound to ROSs in the dark, where all G proteins are in the trimeric (E_s) form. The G_α* or E_h form displays high affinity for GTP just as the E_h form of ATP synthase has a high affinity for ATP.

Another similarity is observed between F_1 and the G protein. In the presence of high concentrations of Mg^{2+} (and under certain experimental circumstances), ADP stabilizes the E_h form of ATP synthase, preventing or abolishing ATP hydrolysis.[43,44] In the presence of high concentrations of Mg^{2+} (>10 mM) and in the absence of a signalling receptor and P_i, the G protein is also unable to hydrolyze the triphosphate, probably due to dissociation of G_α yielding the extraordinary species G_α*·GDP·Mg^{2+} (refs. 23,45), i.e., the E_h form.

Finally, an artificially decoupled (soluble) F_1-unit acts in a similar way as a dissociated G_α* unit. After decoupling from the protonmotive force, both proteins only display hydrolysis activity. The essence of chemiosmotic theory (chapter 1) is that only when a high enough Δp is coupled to $F_1 \cdot F_0$-ATP synthase will this protein actually synthesize ATP; in the absence of a Δp the protein will hydrolyze ATP. This principle also seems to hold for G protein activity: after G_α* dissociation, the G_α(*) subunit is no longer in contact with the Δp provided by its coupling to (H)R*, enabling GTP hydrolysis.

Dissimilarities Between the G Protein and $F_1 \cdot F_0$-ATP Synthase

The G protein differs from F_1 in that the latter cannot release one of its structural units. F_1 possesses six nucleotide binding sites of which three are catalytically active and three are not. The G protein only has one catalytically active nucleotide binding site[46-48] at G_α. The primary sequences of both F_1 and the G protein possess the conserved nucleotide binding motif GxxxxGKT/S (see also chapter 2). In the literature, the possibility of a second nucleotide binding site was suggested,[49-51] although this suggestion is not confirmed by the recently reported X-ray structures of trimeric G proteins[52-53] where only one nucleotide is crystallized together with the protein.

Conversion from the E_h to the E_s Form of G_α

Conversion from the E_s to the E_h form of the G protein involves dissociation of G_α* from the $G_{\beta\gamma}$ dimer, whereas conversion from E_h to E_s involves relaxation from G_α* to G_α (with the possibility of trimer $G_{\alpha\beta\gamma}$ trimer regeneration). Trimer dissociation is accomplished by breaking salt bridges and lipophilic interactions between $G_{\beta\gamma}$ and G_α.[46-54] Modelling studies on the β_1 adrenergic ternary complex have revealed that the conversion from G_α to G_α* could actually be initiated by breaking of the salt bridge that fixes the α_2 helix in the G_α subunit (*vide infra*).[55] Also, cysteine residues and fatty acid coupling might play a role in trimer association.[56] The slow trimer regeneration process[57-58] might thus be due to a time-consuming restoration of salt bridges, lipophilic interactions, disulfide bridges, lipidation states of G_α and $G_{\beta\gamma}$, and/or the phosphorylation state of a histidine residue on the G_β subunit.

8.3 POSTULATE 3: GPCRs correspond to the F_0 unit of ATP synthases

The F_0 unit of ATP synthase is suggested to adapt different conformations depending on whether F_1 is in the synthesis or hydrolysis form (chapter 3). We postulate that GPCRs react in a similar way upon variations in the conformation of G_α. When the G protein in its trimeric E_s form couples to a GPCR, the GPCR is in the R* state with a high affinity for agonists and its hydrogen-bonded chain directed towards the G protein. It has been suggested that part of the HBC consists of residues from the third intracellular loop of the receptor[55]—a conclusion which is partially based on the fact that mutations in this loop are associated with constitutive activation.[25] Samama et al[25] suggest that the third intracellular loop is exposed in the R* form (see also ref. 59), whereas Oliveira c.s.[5] suggest a conserved Arg to be exposed. These suggestions are in accordance with the concept of an HBC inside the receptor making contacts with the HBC inside the G protein.

Taking into consideration that trinitrophenyl-ADP can open the pathway for proton translocation in F_0 by binding to F_1 (chapter 3), one can imagine that binding of the G protein to R* (with the G_α unit of G in the synthesis form) can open or stabilize the hydrogen-bonded chain for proton translocation through the GPCR, leading to the high affinity state for agonists. Modelling studies by our group[55] have shown that this is a tenable hypothesis. Within these studies, Timms and co-workers' proton pump mechanism[31-32] is directly related to G protein activation. The HBC ends in the reaction chamber of the G protein (i.e., the nucleotide binding site together with the phosphorylated histidine at G_β; *vide infra*, postulate 4) explaining why $G_{\alpha\beta\gamma}{\cdot}GDP{\cdot}P_i$ couples to and thus stabilizes R*; in other words, the GDP-bound form of the trimeric G protein has a high affinity for the activated receptor states, R* and HR*. The same modelling studies[55] have also revealed that only R*G or HR*G can consume the Δp. The presence of a G protein is therefore a prerequisite for proton conduction and this also prevents R* or HR* from short-circuiting any Δp. Hence, the activated R* state of the GPCR possesses an HBC inside the central cavity for which both agonists and G proteins have high affinity. But the proton wire inside (H)R* is inactive as long as the G protein does not bind to the appropriate GPCR. In other words, only after the constitution of the ternary complex (or R*G; basal activity) is the proton conductance of the wire such that it can translocate protons from the outside towards the inside of the cell, thereby yielding $G_\alpha^*\cdot GTP$ complexes. It is noted that the molecular mechanisms of agonist binding to the receptor, G protein binding to the receptor and G protein activation are clearly distinct, which is in agreement with Hausdorff and coworkers' observations.[35]

When the G protein is in its hydrolysis form (E_h or G_α^*) and hence dissociated from HR*G or R*G, the GPCR adopts a conformation in which a complete HBC is absent (inducing the inability to translocate protons), corresponding to the low affinity state for agonists denoted by R in the extended ternary complex model.[25] Hence, G protein coupling to the receptor aids in establishing an active HBC by stabilizing R* and is further an absolute necessity for any proton transport from the periplasm toward the cytosol.

8.4 POSTULATE 4: The ternary complex acts as a secondary proton pump and as a GTP synthase

The fourth postulate, closely connected with the previous two, is that both the ternary complex (HR*G) and R*G act as *GTP synthases* using (pre-bound or cytosolic) GDP to synthesize GTP. This is in complete agreement with the earlier mentioned G protein phosphorylation reactions regulated by agonist-activated receptors.[2-3,51,60-61]

THE PUTATIVE ROLE OF CHEMICAL PROTONS IN DEPHOSPHORYLATION

Phosphorylation (or thiophosphorylation) is suggested to proceed via a histidine residue on the G_β subunit.[2-3,51,60-62] It is likely that the histidine is dephosphorylated with the aid of a *chemical* proton. This is inferred from known chemical data on *N'*-phosphorylhistidine, revealing that (covalent) phosphoramidate derivatives, such as phosphohistidine, show acid lability and alkaline stability (ref. 63; and refs. therein). The following reactions are proposed to occur inside the G protein:

$$\text{G-His-PO}_3\text{H} + \text{H}_2\text{O} + \text{H}^+_{(\text{chemical})} \rightleftharpoons \text{G-HisH}^+ + \text{H}_2\text{PO}_4^- \quad (8.1\text{a})$$

$$\text{G-GDP} + H_2PO_4^- \rightleftharpoons \text{G-GTP} + H_2O \qquad (8.1b)$$

The above mentioned chemical proton (Eq. 8.1a) is transferred to the phosphorylated histidine residue on the G_β subunit. Protonation of the imidazole ring system weakens the phosphoramidate binding. Note that within the actual GTP synthesis reaction (Eq. 8.1b) no proton is added on the left hand side, since the proton is implicitly added via $H_2PO_4^-$, i.e., H^+P_i (P_i has a formal charge of -2, cf. Eq. 1.4, chapter 1). Stock et al[64] indicate that transphosphorylation reactions catalyzed by histidine protein kinases occur, in general, via a phosphorylated aspartic acid intermediate, so that a carboxylate group can be expected to be important for the transphosphorylation reaction associated with G protein activation.

If the binding of GDP-P_i to the trimeric form of the G protein is tight, this could result in a situation analogous to the one observed for ATP synthases, where bound ADP and P_i are in equilibrium with bound ATP (K_{eq} ~1) and the actual ATP synthesis step does not require much Gibbs energy input (chapter 3):

$$\text{G-GDP-}P_i \rightleftharpoons \text{G-GTP} \qquad K_{eq} \approx 1 \qquad (8.2)$$

If the nucleotides indeed bind tightly to the G protein, Eq. 8.2 is just a simplified notation of Eq. 8.1 and with the aid of a *chemical proton* the reaction is shifted towards synthesis. The chemical proton can originate from proton translocation via the HR*G complex or from changes in the local environment of the G_β's histidine, causing a pK_a shift in a neighboring residue which leads to the histidine protonation (cf. Eq. 8.1a). Modelling studies may aid in elucidating the exact mechanism.[55]

GTP Synthesis and Hydrolysis Reactions in G Proteins

We thus propose that in the GTP synthesis reaction the (thio)phosphate group bound to the histidine residue in the G_β unit is used and corresponds to the γ-(thio)phosphate of the product. The phosphate will then be stabilized by two positively charged residues in the G_α subunit (Lys[42] and Arg[174] in T-r; refs. 46-47,53). During the dephosphorylation of G_β and subsequent GTP synthesis at G_α, the nucleotide binding site changes to yield the activated G_α* state. In the X-ray structure of T-r_α, the nucleotide binding site is more open than in the activated T-r_α* state, and this open nucleotide binding site at T-r_α hence allows an (indirect) interaction with the aforementioned phosphorylated histidine from the G_β subunit in the trimeric form of T-r. The more open character of the nucleotide binding site in the G_α compared to the G_α* structure corresponds well to the experimental data presented by Robinson and Hagins[65] concerning nucleotide concentration levels. Pre-coupled GDP is embedded in a more open nucleotide binding site and might thus be acid-extractable, whereas GTP bound in a closed cleft could be acid-inextractable. The overall drop observed in guanosine acid-extractable concentration (cf. Fig. 6.1) can then also be explained: acid-extractable GDP (open cleft, G_α) is synthesized into acid-inextractable GTP (closed cleft, G_α*). The tight binding of a triphosphate group within the closed cleft of G_α* could play a role in the synthesis reaction (Eqs. 3.2a-b and 8.2).

Since GTP synthesis occurs in the trimer and GTP hydrolysis occurs at the dissociated G_α* subunit, the phosphorylation and hydrolysis reactions occur at (partially) different sites: i.e., the γ-phosphate is either bound to a histidine residue at the G_β subunit (synthesis) or located in a cleft at the surface of G_α* (hydrolysis). In contrast, the catalysis of ATP synthesis and hydrolysis occurs at the same site of $F_1 \cdot F_0$-ATP synthase. Furthermore, in the case of ATP synthesis no (covalent) phosphoenzyme intermediate has ever been detected, and stereo-inversion at the terminal phosphate is observed. The phosphorylation reaction catalyzed by a G protein (via a covalent phosphoramidate intermediate) is predicted to retain the stereochemistry of the terminal phosphate (cf. chapter 3).

GTP hydrolysis proceeding through a glutamine-activated water molecule[66-67] releases Gibbs free energy, which in turn could be used to open the closed nucleotide cleft as observed in the X-ray structure of T-r_α* associated with a (slow) release of the hydrolysis products (i.e., GDP and P_i) and resulting in the T-r_α from. Since GTPγS cannot be hydrolyzed, no energy is available to release the hydrolysis products, which explains why in general non-hydrolyzable nucleotides induce a continuous activation of the G protein.

In the case of ATP synthases, it is known that nucleotide release costs Gibbs energy.

THE PUTATIVE ROLE OF VECTORIAL PROTONS

T-r_α and T-r_α* differ in three so-called switch regions.[46-47] The three-dimensional structure of two of these regions (I and II) seems to be directly affected by the presence of the γ-phosphate group. Structural changes could also result from the translocation of *vectorial protons*, which could for example be responsible for the disruption of the salt bridge between Glu^{212} and His^{209} (in T-r) and concomitant increase of the rotational freedom of the α_2 helix.[55] The HBC transferring these protons thus runs from the extracellular side of the receptor via intracellular loops to this particular 212-209 salt bridge in T-r_α.

In Figure 8.3 a ligand-β_1 adrenoceptor $G_{s\alpha}$ complex is depicted. The loss of this Glu^{212}-His^{209} connection, together with the introduction of a P_i group at G_α, offers the possibility of increasing the tightness of GDP-P_i binding. This tight binding can facilitate GTP synthesis and again analogies with ATP synthases become apparent: by tightening the GDP-P_i binding (by closing the nucleotide binding site in the G protein) the GTP synthesis reaction requires less Gibbs free energy (Eq. 8.2). Dratz et al (ref. 68, and refs. therein) obtained experimental evidence from NMR studies for the occurrence of conformational changes at the rhodopsin/T-r interface during the G protein activation process. This supports the possibility of an operational hop/turn mechanism at this interface region. Furthermore, it is known that G_α has high affinity for GDP, whereas G_α* has high affinity for GTP. Therefore, the introduction of P_i at G_α during the activation process not only induces the necessary structural changes in switch regions I and II (*vide supra*) but prevents GDP from being released at the same time. This is in close analogy with the role of AlF_4^-, yielding the transition state analagon GDP·AlF_4^- at G_α. Purified G_α reconstituted in the absence of (phosphorylated) $G_{\beta\gamma}$ is known to thrust out its pre-bound GDP when it is activated to its G_α* form (with little affinity for GDP) by an activated GPCR (Misha Freissmuth, personal communication). In this latter case, however, a P_i group is missing to induce structural changes within switches I and II and the

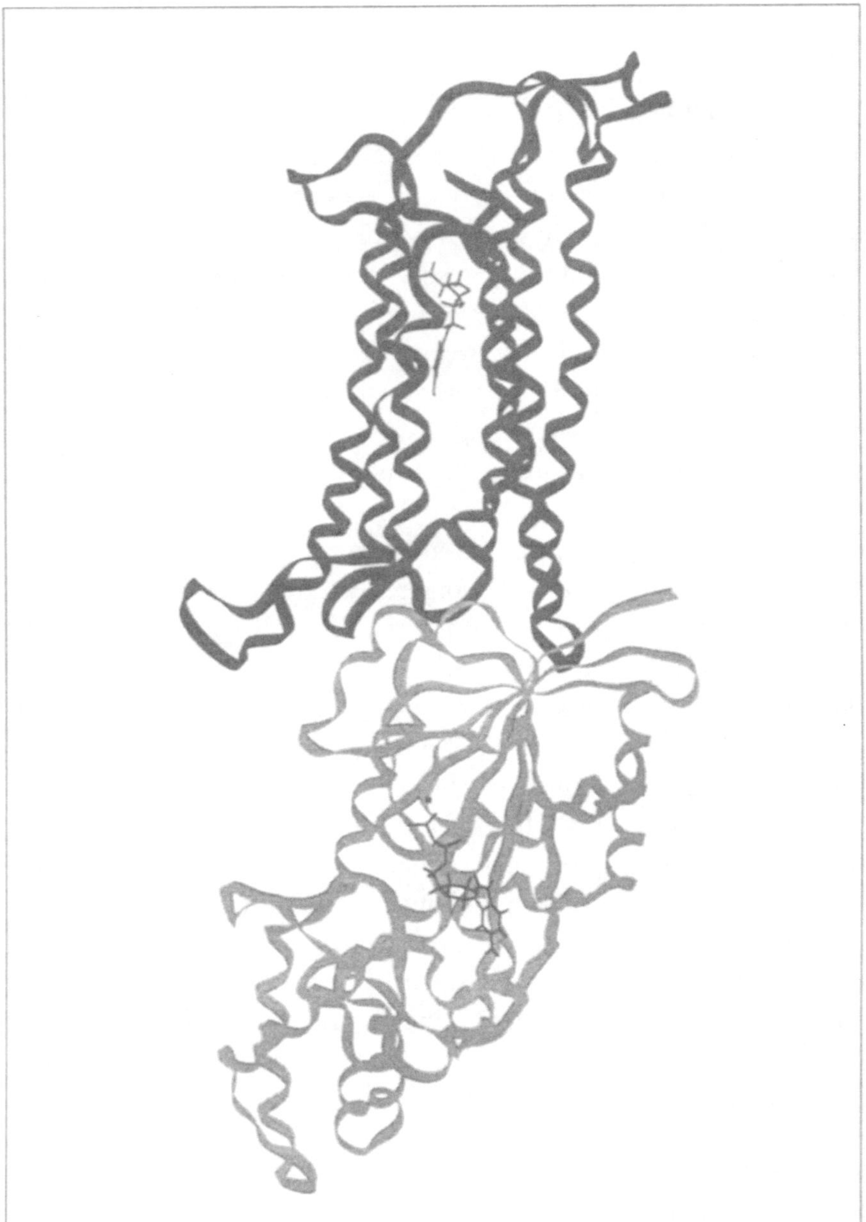

Fig. 8.3. A three-dimensional model of an agonist-β_1 adrenoceptor-$G_{s\alpha}$ complex. The α-carbons of the receptor's seven α-helical proton pump model[31,32] are depicted in dark-gray. The $G_{s\alpha}$ subunit is colored light-gray with the guanine nucleotide and the agonist in black. The relative orientation of the two proteins is partially determined by the interaction of the receptor's HBC with the 212-209 salt bridge at the G_α's α_2 helix (amino acid numbering taken from transducin). For further details the reader is referred to Nederkoorn et al.[55] The figure is rendered with the program SETOR.[74]

pre-bound nucleotide can thus not be prevented from being expelled. Note that R* now acts as a guanine nucleotide exchange factor (GEF) for G_α (cf. chapter 7).

The rotation of the α_2 helix within the G protein and the other changes observed within the switch regions upon G protein activation[46-47] are also responsible for the dissociation of $G_\alpha{}^*$ from $G_{\beta\gamma}$ through the disruption of the α-$\beta\gamma$ interface (chapter 4).[52-54] In our model, the protonmotive force Δp can deliver the energy necessary to convert the ternary complex from the so-called E_s into the E_h form, i.e., to activate the G protein via the translocation of chemical and/or vectorial protons.

Extension of the New Model to Low (Nonphysiological) GTP Levels

In this subsection we aim to explain the stable state of the ternary complex HR*G or R*G observed at GTP concentrations far below the physiolological levels assumed to be present in the dark adapted state of ROSs (ref. 36, and refs. therein).

At the very moment a GPCR is activated from the resting state (the dark-adapted state for ROSs), we propose that the ternary complex starts to synthesize GTP with the aid of a protonmotive force. GTP concentrations are then at a physiological level. The synthesis reaction requires one chemical proton and possibly additional vectorial proton(s), which are used to dephosphorylate a particular histidine residue at G_β and induce dissociation of the $G_\alpha{}^*$·GTP unit. Thus, on activation and at physiological GTP levels both the G protein and the ternary complex are ephemeral.

However, when GTP levels decrease, and in the most extreme test-tube circumstances can even be set to zero, the ternary complex becomes stable, i.e., neither $G_\alpha{}^*$·GTP dissociation nor a labile HR*G is observed. There are two possible explanations for this phenomenon: (i) In our opinion, the most likely option is that in the absence of GTP or at extremely low levels, the G protein cannot be phosphorylated. Experimental evidence for this option can be found in refs. 2-3, 51 and 60-62. If the G_β is not phosphorylated, GTP synthesis cannot occur, which in turn prevents the activation of the G protein. (ii) A second possibility is that at extremely low GTP levels, the high energy intermediate H_2GTP^{2-} is formed and subsequently released into the cytosol to enhance GTP levels; G_α^* does not dissociate. This second option is in line with the ATP synthesis reaction (chapter 1, Eq. 1.6), where ATP synthesis depends on the [ATP]/[ADP] ratio and the height of the protonmotive force. ATP is synthesized and released until the [ATP]/[ADP] ratio becomes too high. For GTP synthesis a similar equation can then be established:

$$GDP + P_i + ESE \rightleftharpoons H_2GTP^{2-} \qquad (8.3)$$

We refer to chapter 6, where we discussed observations pro and contra GTP synthesis at the G protein, showing that GDP can indeed dephosphorylate a phosphorylated G protein, resulting in the formation of free GTP.[60,69]

8.5 POSTULATE 5: Cytosensor microphysiometers measure primary pump activities

The fifth postulate states that the proton outburst following receptor activation, can (partially) be seen as a result from the activity of primary pumps such as sodium/proton antiporters, glycolysis, the actions of protein kinase C, or tyrosine kinases. This proton flux has actually been measured with a cytosensor microphysiometer system (CMS)[70] (CMS is available from Molecular Devices Corp.; ref. 71). Since we consider the ternary complex and R*G as secondary pumps, in other words as GTP synthases which consume Δp, primary pumps are a prerequisite for constituting a proton circuit. The extracellular increase in acidity on GPCR activation as monitored by the CMS proves the existence of primary pumps in the cell. Further evidence is available from experiments reported by Suszták et al,[72] which show that ligands of purinergic G protein-coupled receptors stimulate H^+ transport out of neutrophils. The large cytosolic Na^+ and ATP turnover, found in activated ROSs (cf. 65, and refs. therein), might also provide additional evidence for the presence of primary pumps, establishing a proton gradient.

pH EFFECTS

Since other proton-pumping proteins are likely to be involved in GPCR activation, relating pH changes to GPCR activity is complicated. Furthermore, pH changes following illumination of ROSs might even be absent, as indicated by Stryer.[9] This observation is to be expected when the ternary complex is part of a localized closed circuit[73] in which protons are transferred directly from the GPCR via the membrane interface toward the primary pump(s). Hence, no periplasmic or cytosolic pH changes will result (chapter 1, Figs. 1.1 and 1.3). Only when the proton sink (i.e. the secondary proton pump) is blocked, while the primary pump(s) is still active, will protons no longer diffuse along the membrane, but slowly dissipate into the bulk phase leading to pH effects which can be monitored by the CMS. This situation may occur after depletion of the G protein pool (the proton sink (H)R*G is no longer present for proton translocation, and hence the localized circuit is no longer active).

Partial Agonism

The protonmotive force, Δp, consists of two components (chapter 1), one due to the electrical potential build up over the membrane ($\Delta\psi$) and the other due to a difference in the concentration of protons across the membrane (ΔpH). In our opinion, the influence of both $\Delta\psi$ and of ΔpH should be regarded as being localized, because the proton circuit has been indicated to be closed.[73] Δp depends on the kind of tissue under consideration and the species used. This could explain why partial agonists display species- and tissue-dependent intrinsic activities. Furthermore, the proton circuit of which the GPCR under consideration is part may vary (e.g., different primary pumps) with obvious consequences. In addition, when kinases phosphorylate in a pH dependent manner, variations in Δp might be a source of different desensitization patterns, both in time and effect.

8.6 SUMMARY

Through five postulates bioenergetical principles have been applied to ternary complex and G protein activation. Both agonists and G proteins have been indicated to stabilize the active R* state of a GPCR, which we claim to correspond to the constitution of an HBC inside the receptor's central cavity. G protein coupling is, however, indicated to be essential for proton transport over this proton wire, which explains why activated receptors cannot short-circuit the Δp of a cell. Functional analogies between ternary complexes and $F_1 \cdot F_0$-ATP synthases have been outlined. The introduction of a proton at the phosphorylated histidine at G_β (chemical proton) is held responsible for releasing P_i into the nucleotide binding site at G_α, which not only prevents GDP from being released by inducing the necessary structural changes inside the G_α's switch regions I and II (analogous to the AlF_4^- mechanism) but also plays a key role in the subsequent GTP synthesis reaction. From modelling studies (cf. Fig. 8.3) it is inferred that a vectorial proton can rupture an essential salt bridge at G_α, thereby weakening the inactive G_α structure and enabling the rotation of the α_2 helix necessary to adapt the active G_α^* conformation. Whether or not the vectorial proton(s) and chemical proton are one and the same (i.e. only one proton is translocated) needs to be elucidated; in other words, at this stage nothing can be said about the H^+/GTP ratio. If only one proton per agonist bound is indeed responsible for the activation of one G protein, during the translocation of this proton from the extracellular space towards the phosphorylated histidine on G_β (chemical role), this proton must rupture the aforementioned salt bridge (vectorial role) as well. Each cell possesses primary pumps and some examples have been given. In addition, several possibilities for the experimental verification of proton translocation by ternary complexes have been outlined.

References

1. Topiol S. The deletion model for the origin of receptors. Trends Biochem Sci 1987; 12:419-421.
2. Wieland T, Kaldenberg-Stasch S, Fesseler B et al. Regulation of G protein function by phosphorylation. Can J Physiol Pharmacol 1994; 72:S5.
3. Wieland T, Nürnberg B, Ulibarri I et al. Guanine nucleotide-specific phosphate transfer by guanine nucleotide-binding regulatory protein β-subunits. J Biol Chem 1993; 268:18111-18118.
4. Nederkoorn PHJ, Timmerman H, Donné-Op den Kelder GM. Does the ternary complex act as a secondary proton pump and a GTP synthase? Trends Pharmacol Sci 1995; 16:156-161.
5. Oliveira L, Paiva ACM, Sander C et al. A common step for signal transduction in G protein-coupled receptors. Trends Pharmacol Sci 1994; 15:170-172.

6. Sealfon SC, Chi L, Ebersole BJ et al. Related contribution of specific helix 2 and 7 residues to conformational activation of the serotonin 5-HT_{2A} receptor. J Chem Biol 1995; 270:16683-16688.
7. Dencher NA, Büldt G, Heberle J et al. Light-triggered opening and closing of a hydrophobic gate controls vectorial proton transfer across bacteriorhodopsin. NATO ASI Ser, Ser B 1992; 291:171-185.
8. Hofmann KP. Photoproducts of rhodopsin in the disc membrane. Photobiochem Photobiophys 1986; 13:309-327.
9. Stryer L. Cyclic GMP cascade of vision. Ann Rev Neurosci 1986; 9:87-119.
10. Findlay JBC, Pappin DJC. The opsin family of proteins. Biochem J 1986; 238:625-642.
11. Oliveira L, Paiva ACM, Vriend G. A common motif in G protein-coupled seven transmembrane helix receptors. J Comp-Aided Mol Design 1993; 7:649-658.
12. Sakmar TP, Franke RR, Khorana HG. Mutagenesis studies of rhodopsin phototransduction. In: Hargrave P, ed. Signal transduction in photoreceptor cells. Berlin: Springer Verlag, 1992:21-30.
13. Deng H, Huang L, Callender R et al. Evidence for a bound water molecule next to the retinal Schiff base in bacteriorhodopsin and rhodopsin: A resonance Raman study of the Schiff base hydrogen/deuterium exchange. Biophys J 1994; 66:1129-1136.
14. Arnis S, Hofmann KP. Two different forms of metarhodopsin II: Schiff base deprotonation precedes proton uptake and signalling state. Proc Natl Acad Sci USA 1993; 90:7849-7853.
15. Schleicher A, Hofmann KP. Proton uptake by light-induced interaction between rhodopsin and G-protein. Z Naturforsch Teil C 1985; 40:400-405.
16. Spudich JL. Protein-protein interaction converts a proton pump into a sensory receptor. Cell 1994; 79:747-750.
17. Higashijima T, Burnier J, Ross EM. Regulation of G_i and G_o by mastoparan, related amphiphilic peptides, and hydrophobic amines. Mechanism and structural determinants of activity. J Biol Chem 1990; 265:14176-14186.
18. Higashijima T, Uzu S, Nakajima T et al. Mastoparan, a peptide toxin from wasp venom, mimics receptors by activating GTP-binding regulatory proteins (G proteins). J Biol Chem 1988; 263:6491-6494.
19. König B, Arendt A, McDowell JH et al. Three cytoplasmic loops of rhodopsin interact with transducin. Proc Natl Acad Sci USA 1989; 86:6878-6882.
20. Münch G, Dees C, Hekman M et al. Multisite contacts involved in coupling of the β-adrenergic receptor with stimulatory guanine nucleotide-binding regulatory protein. Structural and functional

studies by β-receptor-site-specific synthetic peptides. Eur J Biochem 1991; 198:357-364.

21. Okamoto T, Murayama Y, Hayashi Y et al. Identification of the β_2 adrenergic receptor that is autoregulated via protein kinase A-dependent phosphorylation. Cell 1991; 76:723-730.
22. Ross EM. G protein-coupled receptors: structural basis of selective signalling. NATO ASI Series, Series H 1991; 52:163-177.
23. Birnbaumer L, Birnbaumer M. Signal transduction by G proteins: 1994 edition. J Receptor & Signal Transduction Research 1995; 15:213-252.
24. Keefer JR, Nunnari J, Pang IH et al. Introduction of purified α_{2A} adrenergic receptors into uniformly oriented, unilamellar, phospholipid vesicles: productive coupling to G proteins but lack of receptor-dependent ion transport. Mol Pharmacol 1994; 45:1071-1081.
25. Samama P, Cotecchia S, Costa T et al. A mutation-induced activated state of the β_2 adrenergic receptor. Extending the ternary complex model. J Biol Chem 1993; 268:4625-4636.
26. Nicholls DG, Ferguson SJ. In: Bioenergetics 2. London: Academic Press, 1992.
27. Freer JH, Arbuthnott JP. Toxins of Staphylococcus aureus. Pharmac Ther 1983; 19:55-106.
28. Kaldenberg-Stasch S, Liedel K, Hagelskamp E et al. Analysis of receptor-G protein interactions in permeabilized cells. Naunyn-Schmiedeberg's Arch Pharmacol 1995; 351:R71.
29. Wikström M. Proton pump coupled to cytochrome *c* oxidase in mitochondria. Nature 1977; 266:271-273.
30. Rigaud J-L, Paternostre M-T, Bluzat A. Mechanisms of membrane protein insertion into liposomes during reconstitution procedures involving the use of detergens. 2. Incorporation of the light-driven proton pump bacteriorhodopsin. Biochemistry 1988; 27:2677-2688.
31. Timms D, Wilkinson AJ, Kelly DR et al. Interactions of Tyr^{377} in a ligand-activation model of signal transmission through β_1 adrenoceptor α-helices. Int J Quant Chem: Quant Biol Symp 1992; 19:197-215.
32. Timms D,Wilkinson AJ, Kelly DR et al. Ligand-activated transmembrane proton transfer in β_1 adrenergic and m_2 muscarinic receptors. Receptors and Channels 1994; 2:107-119.
33. Schreuder HA, Mattevi A, Obmolova G et al. Crystal structures of wild-type p-hydroxybenzoate hydroxylase complexed with 4-aminobenzoate and of the tyr222ala mutant complexed with 2-hydroxy-4-aminobenzoate. Evidence for a proton channel and a new binding mode of the flavin ring. Biochemistry 1994; 33:10161-10170.

34. van der Goot H, Bast A, Timmerman H. Structural requirements for histamine H_2 agonists and H_2 antagonists. In: Uvnäs B, ed. Handbook of experimental pharmacology, Vol 97. Berlin: Springer Verlag, 1991:573-748.
35. Hausdorff WP, Hnatowich M, O'Dowd BF et al. A mutation of the β_2 adrenergic receptor impairs agonist activation of adenylate cyclase without affecting high affinity agonist binding. J Biol Chem 1990; 265:1388-1393.
36. Schütz W, Freissmuth M. Reverse intrinsic activity of antagonists on G-protein-coupled receptors. Trends Pharmacol Sci 1992; 13:376-379.
37. Lefkowitz RJ, Cotecchia S, Samama P et al. Constitutive activity of receptors coupled to guanine nucleotide regulatory proteins. Trends Pharmacol Sci 1993; 14:303-307.
38. Milligan G, Bond RA, Lee M. Inverse agonism: pharmacological curiosity or potential therapeutic strategy? Trends Pharmacol Sci 1995; 16:10-13.
39. Strader CD, Dixon RAF, Cheung AH et al. Mutations that uncouple the beta-adrenergic receptor from G_s and increase agonist affinity. J Biol Chem 1987; 262:16439-16443.
40. Ernst OP, Hofmann KP, Sakmar TP. Characterization of rhodopsin mutants that bind transducin but fail to induce GTP nucleotide uptake. J Biol Chem 1995; 270:10580-10586.
41a. Leurs R, Smit MJ, Tensen CP et al. Site-directed mutagenesis of histamine H_1-receptor reveals a selective interaction of asparagine[207] with subclasses of H_1 receptor agonists. Biochem Biophys Res Comm 1994; 201:295-301.
41b. Leurs R, Smit MJ, Meerder R et al. Lysine[200] located in the fifth transmembrane domain of the histamine H_1 receptor interacts with histamine but not all H_1 agonists. Biochem Biophys Res Comm 1995; 214:110-117.
41c. ter Laak AM, Timmerman H, Leurs R et al. Modelling and mutations studies on the histamine H_1 receptor agonist binding site reveal different binding modes for H_1 agonists; Asp[116] (TM3) has a constitutive role in receptor stimulation. J Comp-Aided Mol Design 1995; 9:319-330.
42. Wiens BL, Nichols DE, Mailman RB et al. Multiple determinants of agonist efficacy at dopamine D_2 receptors. Soc Neurosci Abstracts 1994; 20:524.
43. Boyer PD. A perspective of the binding change mechanism for ATP synthesis. FASEB J 1989; 3:2164-2178.
44. Guerrero KJ, Xue Z, Stempel KE et al. Further insight into the regulation of CF_1-ATPase activity by ADP and Mg^{2+}. J Cell Biol 1988; 107:627a.
45. Gilman AG. G proteins: Transducers of receptor-generated signals. Annu Rev Biochem 1987; 56:615-649.

46. Noel JP, Hamm HE, Sigler PB. The 2.2 Å crystal structure of transducin-α complexed with GTPγS. Nature 1993; 366:654-663.
47. Lambright DG, Noel JP, Hamm HE et al. Structural determinants for activation of the α-subunit of a heterotrimeric G protein. Nature 1994; 369:621-628.
48. Coleman DE, Berghuis AM, Lee E et al. Structures of active conformations of $G_{i\alpha 1}$ and the mechanism of GTP hydrolysis. Science 1994; 265:1405-1412.
49. Fung BK-K, Stryer L. Photolyzed rhodopsin catalyzes the exchange of GTP for bound GDP in retinal rod outer segments. Proc Natl Acad Sci USA 1980; 77:2500-2504.
50. Godchaux W III, Zimmerman WF. Membrane-dependent guanine nucleotide binding and GTPase activities of soluble protein from bovine rod cell outer segments. J Biol Chem 1979; 254:7874-7884.
51. Kaldenberg-Stasch S, Baden M, Fesseler B et al. Receptor-stimulated guanine nucleotide-triphosphate binding to guanine nucleotide-binding regulatory proteins. Eur J Biochem 1994; 221:25-33.
52. Wall MA, Coleman DE, Lee E et al. The structure of the G protein heterotrimer $G_{i\alpha 1\beta 1\gamma 2}$. Cell 1995; 83:1047-1058.
53. Lambright DG, Sondek J, Bohm A et al. The 2.0 Å crystal structure of a heterotrimeric G protein. Nature 1996; 379:311-319.
54. Sondek J, Bohm A, Lambright DG et al. Crystal structure of a G_A protein βγ dimer at 2.1Å resolution. Nature 1996; 379:369-374.
55. Nederkoorn PHJ, Timmerman H, Donné-Op den Kelder GM et al. GTP synthases. Proton pumping and phosphorylation in G_α protein-receptor complexes. Receptors & Channels 1996; 4:111-128.
56. Casey PJ. Protein lipidation in cell signalling. Science 1995; 268:221-225.
57. Chabre M, Antonny B, Vuong TM. The transducin cycle in the phototransduction cascade. NATO ASI Series, Series H 1991; 52:207-220.
58. Fung BK-K. Charaterization of transducin from bovine retinal rod outer segments. Separation and reconstitution of the subunits. J Biol Chem 1983; 258:10495-502.
59. Weinstein H. Computational simulations of molecular structure, dynamics and signal transduction in biological systems: Mechanistic implications for ecological physical chemistry. In: Bonati L, Cosentino U, Lasagni M et al, eds. Trends in ecological physical chemistry, Proceedings of the 2nd international workshop on ecological physical chemistry. Amsterdam: Elsevier 1993:1-16.
60. Kowluru A, Seavey SE, Rhodes CJ et al. A novel regulatory mechanism for trimeric GTP-binding proteins in the membrane and secretory granule fractions of human and rodent β cells. Biochem J 1996; 313:97-107.

61. Seifert R. Transducin is a nucleoside diphosphate kinase. Naunyn-Schmiedeberg's Arch Pharmacol 1995; 351:R68.
62. Klinker JF. Transducin possesses NTPase activity. Naunyn-Schmiedeberg's Arch Pharmacol 1995; 351:R68.
63. VanEtten RL, Waymack PP, Rehkop DM. Transition metal ion inhibition of enzyme-catalyzed phosphate ester displacement reactions. J Am Chem Soc 1974; 96:6782-6785.
64. Stock JB, Ninfa AJ, Stock AM. Protein phosphorylation and regulation of adaptive responses in bacteria. Microbiol Rev 1989; 53:450-490.
65. Robinson WE, Hagins WA. GTP hydrolysis in intact rod outer segments and the transmitter cycle in visual excitation. Nature 1979; 280:398-400.
66. Kleuss C, Raw AS, Lee E et al. Mechanism of GTP hydrolysis by G protein α subunits. Proc Natl Acad Sci USA 1994; 91:9829-9831.
67. Sondek J, Lambright DG, Noel JP et al. GTPase mechanism of G proteins from the 1.7 Å crystal structure of transducin $\alpha \cdot GDP \cdot AlF_4^-$. Nature 1994; 372:276-279.
68. Dratz EA, Furstenau JE, Lambert CG et al. NMR structure of a receptor-bound G protein peptide. Nature 1993; 363:276-281.
69. Hohenegger M, Mitterauer T, Voss T et al. Thiophosphorylation of the G protein β subunit in human platelet membranes: evidence against a direct phosphate transfer reaction to G_α subunits. Mol Pharmacol 1996; 49:73-80.
70. McConnell HM, Owicki JC, Parce JW et al. The cytosensor microphysiometer: biological applications of silicon technology. Science 1992; 257:1906-1912.
71. The cytosensor microphysiometer system can be obtained from Molecular Devices Corporation, 1311 Orleans Drive, Sunnyvale, CA 94089, USA
72. Szuszták K, Káldi K, Kapus A et al. Ligands of purinergic receptors stimulate electrogenic H^+-transport of neutrophils. FEBS Lett 1995; 375:79-82.
73. Heberle J, Riesle J, Thiedemann G et al. Proton migration along the membrane surface and retarded surface to bulk transfer. Nature 1994; 370:379-382.
74. Evans SV. SETOR: hardware lighted by three-dimensional solid model representations of macromolecules. J Mol Graphics 1993; 11:134-138.

CHAPTER 9

Conclusions and Future Perspectives

We have summarized those bioenergetical processes which we applied to GPCR signalling, and presented a new model for transmembrane signal transduction mediated via (ligand-induced) GPCR and G protein activation. The classical model for G protein activation is based on an exchange mechanism followed by GTP hydrolysis. Our new model is based on GTP synthesis from GDP and P_i, also followed by GTP hydrolysis. Because of amplification levels (1) (synthesis) and (3) (exchange) depicted in Figure 7.1, both the classical and our new model will give rise to similar nucleotide exchange patterns.

Within the GTP synthase concept, we are able to provide explanations for observations that challenge the present-day model for G protein activation. In particular, our model explains the large drop in GDP levels immediately after activation and the hardly changing level of total acid-extractable GTP in the same time period (0-4 s). Also (thio)phosphorylation reactions mediated via the activated ternary complex can now be put into context. The first step in our activation mechanism (GTP synthesis) is in full agreement with the reported phosphorylation reactions regulated by G proteins.[1-4]

Since signal transduction through GPCRs is considered to occur via proton wires (consisting of a complicated network of hydrogen bonds including waters of hydration), results of mutation studies are less easy to interpret within the new concept. A mutation can disturb the proton wire directly when an amino acid which is part of the HBC is altered. However, when a mutation disturbs the coupling of waters of hydration, it can also lead to a less effective HBC, which in turn will result in a decrease of agonist affinity and/or disturbed proton conductance. This implies, among other things, that the intrinsic activity can be affected, although the agonistic binding site could be far removed from the actual mutation site. When mutations directly destroy the HBC, the chain can be restored (although less effectively) by waters of hydration. Finally, agonists could compete with waters for binding to the HBC, possibly at various sites. This means that different agonistic binding sites can be present on the HBC, while the various agonists expel waters of hydration from different locations on the HBC. The existence of different agonistic binding sites on one and the same GPCR has recently been validated by independent experiments.[5a-c,6]

The essential polar character of the receptor cavity is now put in a broader context, since hop/turn proton translocation mechanisms can only occur in a polar environment. Mutations of amino acids in GPCRs that do not influence agonist binding significantly, but do reduce the overall activity at the second messenger level, can assist in elucidating the position of the alleged HBC. Examples of such mutations have been discussed in chapter 5. It was concluded that the molecular mechanisms for G protein binding to a GPCR and G protein activation are distinct.[7,8] In the extended ternary complex model (ETCM), a particular conformational change (R => R*) is claimed to underlie G protein activation.[9] In other words, the amount of G protein binding to the receptor population with the active conformation R* is directly related to the net rate of G protein activation. In the light of the experimental data reported by Hausdorff and co-workers and Ernst and co-workers[7,8] revealing that a certain mutated GPCR retained both G protein binding and high affinity agonist binding, yet in combination with induced impaired G protein activation, the direct relationship between G protein coupling to R* and G protein activation might be an oversimplification. Within our proton wire concept, we clearly discriminate between the process of G protein coupling to a GPCR and the signal transduction mechanisms (proton translocation from the periplasm towards the cytosol) which lie at the basis of the actual G protein activation.

In the underlying theoretical investigation we have revealed similarities between bacteriorhodopsin (BR) and rhodopsin concerning their proton transfer mechanisms, relative positions of acidic residues and the lysine binding retinal, while waters of hydration seem to be involved in the activation mechanism of both proteins. In addition, Ross[10] points to a possible crucial role for a carboxylate group present in rhodopsin and suggests this residue to have a similar function to that of the highly conserved Asp present in the third transmembrane domain of aminergic GPCRs. Possibly, both aspartates fulfill a crucial role in the proton translocating process, just like an aspartate in the c-chain of the F_0 unit in ATP synthases (chapter 2). It is noted that within Timms' proton pumping model, this Asp is not directly involved.[11,12]

Several authors indicate that the challenge for the future is to discover how the intracellular domains of activated GPCRs bind to the heterotrimer and how the activation of the G protein is actually catalyzed.[13,14] For "theoreticians" many possibilities for further investigations emerge from our new models such as elucidating the exact position of the HBCs within the different GPCRs and trying to model mechanisms for proton translocation and G protein activation. This offers the possibility of quantifying signal transduction processes. We hope that "experimentalists" from the fields of both pharmacology and bioenergetics will be challenged to test our postulates and check our suggestions through experiments. Experimental set-ups with only one type of purified GPCR and G protein, together with controllable primary pumps, reconstituted into lipid vesicles, would provide an excellent basis for studying proton transfers and G protein activation. By controlling the primary pump activity (with light for BR, and with O_2 concentration for cytochrome *c* oxidases, respectively), the Δp can be regulated. The influence of the Δp on GPCR activation processes can then be addressed directly.

The presence of a Δp might further be important for a correct insertion of the reconstituted GPCR, since in the presence of a Δp positively charged amino acids are found preferentially at the intracellular side of membranes (so-called 'positive inside' rule).[15,16] Receptor up/downregulation processes observed in intact cells could very well be related to this 'positive inside' rule. It is important to realize that in intact cells many proton pumping proteins are present and that one must be very careful in relating proton translocation processes monitored along the membrane and pH changes directly to GPCR activity. We finally speculate that new classes of drugs modifying proton translocation at the level of GPCRs as well as at the level of G proteins could emerge. Seifert et al[17,18] already claim that some histamine H_1 receptor agonists (such as 2-(3-chlorophenyl)histamine), by analogy with other cationic-amphiphilic substances, activate the G protein directly, i.e., without any receptor interference.

To speak in Morowitz's terms:[19] The aim of our study is largely heuristic, directed at examining the theoretical possibilities that emerge from interfacing the chemiosmotic views and signal transduction via G protein-coupled receptors. This synthesis produces a number of challenges to experimentalists and hopefully stimulates a series of experiments. *Our postulates are now open for falsification.*

Note Added in Proof

In section 3.3, we outlined the possibility of a rotational movement of the catalytic sites on F_1 of $F_1 \cdot F_0$-ATP synthase. By applying polarized absorption relaxation after photobleaching eosin-labelled F_1-ATPase (the lable was attached to the γ subunit) Sabbert, Engelbrecht and Junge (Nature 13 June 1996; 381:623-625) indeed observed intersubunit rotation in the active F_1-ATPase CF_1 from spinach chloroplast on ATP hydrolysis. The γ subunit rotates over approximately 200° around its long axis within and relative to the hexagon of the $\alpha\beta$ subunits that contain the catalytic nucleotide binding sites (cf. Fig. 1.1). The rate of the rotation complies with the rate of the ATP hydrolysis. How proton translocation via F_0 generates the torque to rotate the γ subunit within $(\alpha\beta)3$ of F_1 remains to be established.

References

1. Kaldenberg-Stasch S, Baden M, Fesseler B et al. Receptor-stimulated guanine nucleotide-triphosphate binding to guanine nucleotide-binding regulatory proteins. Eur J Biochem 1994; 221:25-33.
2. Wieland T, Kaldenberg-Stasch S, Fesseler B et al. Regulation of G protein function by phosphorylation. Can J Physiol Pharmacol 1994; 72:S5.

3. Wieland T, Nürnberg B, Ulibarri I et al. Guanine nucleotide-specific phosphate transfer by guanine nucleotide-binding regulatory protein β-subunits. J Biol Chem 1993; 268:18111-18118.
4. Kowluru A, Seavey SE, Rhodes CJ et al. A novel regulatory mechanism for trimeric GTP-binding proteins in the membrane and secretory granule fractions of human and rodent β cells. Biochem J 1996; 313:97-107.
5a. Leurs R, Smit MJ, Tensen CP et al. Site-directed mutagenesis of histamine H_1-receptor reveals a selective interaction of asparagine[207] with subclasses of H_1 receptor agonists. Biochem Biophys Res Comm 1994; 201:295-301.
5b. Leurs R, Smit MJ, Meerder R et al. Lysine[200] located in the fifth transmembrane domain of the histamine H_1 receptor interacts with histamine but not all H_1 agonists. Biochem Biophys Res Comm 1995; 214:110-117.
5c. ter Laak AM, Timmerman H, Leurs R et al. Modelling and mutations studies on the histamine H_1 receptor agonist binding site reveal different binding modes for H_1 agonists; Asp[116] (TM3) has a constitutive role in receptor stimulation. J Comp-Aided Mol Design 1995; 9:319-330.
6. Wiens BL, Nichols DE, Mailman RB et al. Multiple determinants of agonist efficacy at dopamine D_2 receptors. Soc Neurosci Abstracts 1994; 20:524.
7. Hausdorff WP, Hnatowich M, O'Dowd BF et al. A mutation of the β_2 adrenergic receptor impairs agonist activation of adenylate cyclase without affecting high affinity agonist binding. J Biol Chem 1990; 265:1388-1393.
8. Ernst OP, Hofmann KP, Sakmar TP. Characterization of rhodopsin mutants that bind transducin but fail to induce GTP nucleotide uptake. J Biol Chem 1995; 270:10580-10586.
9. Samama P, Cotecchia S, Costa T et al. A mutation-induced activated state of the β_2 adrenergic receptor. Extending the ternary complex model. J Biol Chem 1993; 268:4625-4636.
10. Ross EM. G protein-coupled receptors: structural basis of selective signalling. NATO ASI Series, Series H 1991; 52:163-177.
11. Timms D, Wilkinson AJ, Kelly DR et al. Interactions of Tyr[377] in a ligand-activation model of signal transmission through β_1 adrenoceptor α-helices. Int J Quant Chem: Quant Biol Symp 1992; 19:197-215.
12. Timms D,Wilkinson AJ, Kelly DR et al. Ligand-activated transmembrane proton transfer in β_1 adrenergic and m_2 muscarinic receptors. Receptors and Channels 1994; 2:107-119.
13. Coleman DE, Sprang SR. How G proteins work: a continuing story Trends Biochem Sci 1996; 21:41-44.

14. Clapham DE. The G protein nanomachine. Nature 1996; 379:297-299.
15. Andersson H, von Heijne G. Membrane protein topology effects of $\Delta_{\mu H^+}$ on the translocation of charged residues explain the 'positive inside' rule. EMBO J 1994; 13:2267-2272.
16. von Heijne G. The distribution of positively charged residues in bacterial inner membrane proteins correlates with the trans-membrane topology. EMBO J 1986; 5:3021-3027.
17. Seifert R, Hagelücken A, Höer A et al. The H_1 receptor agonist 2-(3-chlorophenyl)histamine activates G_i proteins in HL-60 cells through a mechanism that is independent of known histamine receptor subtypes. Mol Pharmacol 1994; 45:578-586.
18. Hagelücken A, Grünbaum L, Klinker JF et al. Histamine receptor-dependent and/or -independent activation of guanine nucleotide-binding proteins by histamine and 2-substituted histamine derivatives in human leukemia (HL-60) and human erythroleukemia (HEL) cells. Biochem Pharmacol 1995; 49:901-914.
19. Morowitz HJ. Proton semiconductors and energy transduction in biological systems. Am J Physiol 1978; 235:R99-R114.

INDEX

U

V

W

X

SPRINGER NATURE

GPSR Compliance

The European Union's (EU) General Product Safety Regulation (GPSR) is a set of rules that requires consumer products to be safe and our obligations to ensure this.

If you have any concerns about our products, you can contact us on ProductSafety@springernature.com

In case Publisher is established outside the EU, the EU authorized representative is:

Springer Nature Customer Service Center GmbH
Europaplatz 3
69115 Heidelberg, Germany

FSC
www.fsc.org
MIX
Papier | Fördert gute Waldnutzung
FSC® C083411

Kolibri 360 GmbH
Ferdinand-Jühlke-Straße 7
99095 Erfurt, Deutschland
produktsicherheit@kolibri360.de